Environmental Science and Engineering

Environmental Science

Series Editors

Ulrich Förstner, Technical University of Hamburg-Harburg, Hamburg, Germany

Wim H. Rulkens, Department of Environmental Technology, Wageningen, The Netherlands

Wim Salomons, Institute for Environmental Studies, University of Amsterdam, Haren, The Netherlands

The protection of our environment is one of the most important challenges facing today's society. At the focus of efforts to solve environmental problems are strategies to determine the actual damage, to manage problems in a viable manner, and to provide technical protection. Similar to the companion subseries Environmental Engineering, Environmental Science reports the newest results of research. The subjects covered include: air pollution; water and soil pollution; renaturation of rivers; lakes and wet areas; biological ecological; and geochemical evaluation of larger regions undergoing rehabilitation; avoidance of environmental damage. The newest research results are presented in concise presentations written in easy to understand language, ready to be put into practice.

More information about this subseries at http://www.springer.com/series/3234

Sunil Nautiyal · Mrinalini Goswami ·
Puneeth Shivakumar

Field Margin Vegetation and Socio-Ecological Environment

Structural, Functional and Spatio-temporal
Dynamics in Rural-urban Interface
of Bengaluru

 Springer

Sunil Nautiyal (iD)
Centre for Ecological Economics
and Natural Resources
Institute for Social and Economic Change
(ISEC)
Bengaluru, Karnataka, India

Mrinalini Goswami
Centre for Ecological Economics
and Natural Resources
Institute for Social and Economic Change
(ISEC)
Bengaluru, Karnataka, India

Puneeth Shivakumar
Centre for Ecological Economics
and Natural Resources
Institute for Social and Economic Change
(ISEC)
Bengaluru, Karnataka, India

ISSN 1863-5520   ISSN 1863-5539   (electronic)
Environmental Science and Engineering
ISSN 1431-6250   ISSN 2661-8222   (electronic)
Environmental Science
ISBN 978-3-030-69203-2   ISBN 978-3-030-69201-8   (eBook)
https://doi.org/10.1007/978-3-030-69201-8

This Springer imprint is published by the registered company Springer Nature Switzerland AG
The registered company address is: Gewerbestrasse 11, 6330 Cham, Switzerland

Foreword

The world population is currently crossing 7.7 billion and is projected to grow to 9.5 billion by 2050 and more than 12 billion by the end of the twenty-first century. Most of such increase in human population accompanied by livestock population is expected to occur in economically developing nations. Consequently, food and fodder demand will also increase. Agricultural intensification through monoculture-based cropping systems is not a promising strategy to fulfil the future needs as adverse environmental effects that were reported to be unwarranted outcome of such strategy. Conversion of natural and semi-natural habitats to arable farms, increased chemical inputs are among the threats to sustainable agriculture. Agricultural intensification has replaced much of the native vegetation across the world, and it is estimated about 70% of land is under agriculture and/or pasture modified systems in tropical region. Intensive agricultural systems are associated with negative environmental impacts, including decreased biodiversity of wild plants and other associated biota. This can lead to increased pest damage as a result of decline in natural pest control often caused by increased chemical inputs. Various approaches, such as adoption of intercropping, can be taken to mitigate these impacts. However, focussing only on field manipulation might not be sufficient to increase biodiversity on the farmland. Supplementation of such interventions with proper management of the field margins is also required. Field margin vegetation may represent the key semi-natural habitat available to enhance biodiversity in each of the ecoregion. Field margin abundance, location and management practices can determine the environmental benefits obtained by the practicing society.

Past scientific studies have reported on the potential of field margins for food provisioning, overwintering sites and hosts to various predators and parasitoids for enhanced biological control services in agro-ecosystems. The importance of field margin management in arable fields for the provision of foraging habitats, nesting sites, food resources and shelter for invertebrates and vertebrates are the important features of such studies. These benefits can be particularly important after disturbances caused by agricultural practices like tillage, pesticide application and harvesting. Field margin establishment and management is one of the affordable measures by a majority of the farmers due to the associated multiple benefits including biodiversity conservation and ecosystem function values. Understanding the various benefits of field margin and non-crop vegetation in agriculture and the environment is particularly important for designing proper management strategy. Broadly, field margins can be grouped under two major categories: cropped field margins and un-cropped field margins. Cropped field margins contain sown arable crops that are identified using ecological and conservation principles. Margins can be managed using the existing field operations where the cultivated strip land is left to regenerate naturally or planting strips to provide food resources to insects. Un-cropped field margins are set aside margins that are sown (with wild seed mixtures) or left to regenerate naturally without human manipulation. Both cropped and un-cropped field margins can be maintained in various ways including cutting to reduce shading and invasion to the field. Field margins may provide various environmental benefits depending on the establishment and management method employed. Multiple benefits may be achieved where different margin types are incorporated at the same farm because no single field margin is capable of providing the required food and habitat resources to all plants and animal groups. Establishment and management method employed upon the field margin in arable farmland may significantly influence the long-term conservation values. Therefore, the intention of integrating agronomic and biodiversity objectives may be achieved through field margin establishment and management.

The present volume broadly deals with these concepts by characterizing the field margins existing in a transect of urban to peri-urban landscape and assessing the functional value of such diversity. The volume is an important addition to the growing scientific evidence on the importance of managing field margin vegetation to maintain biodiversity for conservation and optimizing the ecosystem functions.

February 2021   K. S. Rao
                                              Professor, University of Delhi
                                                       New Delhi, India
                                          Webpage: people.du.ac.in/~ksrao

Preface

Agricultural landscape development requires prioritization of the issues related to conservation of agrobiodiversity and resource sustainability. In-depth research on conserving agrobiodiversity in changing landscape of India is one of the crucial subjects that requires immediate attention. The research in this endeavor would help in developing appropriate strategies for sustainable development of the landscapes dominated by agriculture. Agrobiodiversity plays major role in improving economy of the farmer, mitigating crop yield losses, conserving the local environment and providing habitats to large number of organisms that benefit the landscapes. Field margin vegetation (FMV) is one of the important components of the agroecosystem which is not studied in greater detail in the context of India.

FMVs, the non-crop floral diversity on the boundary, have diverse functions in supporting ecosystem and society. The book broadly has the foundation built upon case studies conducted in periphery of the city of Bengaluru, India. The chapters of the book have discussed structural and functional attributes of FMV in spatio-temporal patterns along an experimental transect of northern periphery of Bengaluru city. The study has been carried out integrating various techniques with multi-disciplinary methodologies. The study endeavors to arrive at policy pointers based on the research outcomes which are very much relevant for the changing landscape of rural–urban interface.

This book is based on findings from different approaches undertaken to study agroecosystems which has successfully assessed the socioecological characteristics of FMV and their relation with transitional agroecosystems. It has documentation of the species on field boundaries (FMV) and agricultural crops which have been driving changes in FMVs that have immense ecological importance. The adverse impacts of cultivation of new market-driven cash crops have been enumerated in the discussion. The outcomes provide insights to an imperative entity of agroecosystems which was not much accounted and acknowledged separately prior to this research in India.

The findings concluded that there have been structural and functional changes of FMV across the transect. Semi-natural field margins with planted economically important species mixed with natural vegetation, which provide significant economic benefits in terms of fodder, fuel, fibre and food along with providing ecological functions, have been replaced by economically important exotic species. Change in crops, more specifically large-scale plantation of *Eucalyptus* spp., grapes, lawn grass and horticulture crops, has also resulted in diminishing FMV diversity. It is a pertinent need of the hour to strategize conservation of FMVs with optimum economic and ecological benefits. Interventions in vegetative components of naturally regenerated filed margin can increase its potential for agricultural landscape sustainability. The book provides a holistic view of spatio-temporal change of FMVs in the rural–urban interface and has identified the significant role of vegetation on field margins, which is first of its kind in India. Further in-depth research on ecological and economic value of FMV should be carried out so that appropriate solutions can be provided for sustainable socio-ecological development of the landscape dominated by agriculture.

We extend our sincere thanks to the Department of Biotechnology (DBT), Government of India, for giving us the opportunity and the financial support to undertake the study on "Spatial patterns of crop and field boundary vegetation's structural and functional parameters change over time along the experimental transects." This work has been carried out as part of the collaborative Indo-German Research Unit FOR2432 Social-Ecological Systems in the Indian Rural–Urban Interface: Functions, Scales, and Dynamics of Transition (DFG); and The Rural–Urban Interface of Bengaluru—A Space of Transitions in Agriculture, Economics, and Society (DBT).

We are deeply thankful to Prof. K. B. Umesh, Department of Agriculture Economics, GKVK, Bengaluru; Prof. Andreas Buerkert and Dr. Ellenn Hoffman, Organic Plant Production and Agroecosystems Research in the Tropics and Subtropics, Universität Kassel, Witzenhausen, Germany; Prof. Dr. Michael Wachendorf, Dr. Thomas Astor, Grassland Science and Renewable Plant Resources, Organic Agricultural Sciences, Universität Kassel, Germany; Prof. B. V. Chinnappa Reddy, GKVK, Bengaluru; Prof. K. N. Ganeshaiah, former Professor GKVK, Bengaluru, for inputs, suggestions, comments from time to time and co-operation during the entire period of research. We acknowledge the generous support provided for conducting our research by the Directors of Institute for Social and Economic Change, Bengaluru and we are deeply thankful to them. We are thankful to Dr. Rama Rao Nidamamuri (IIST, Thiruvananthapuram) collaborator in this research for the technical support, suggestions, co-operation during the entire period of research. We would like to thank Mr. David Andrés Bernal Hoyo, PhD student at CEENR, ISEC from El Colegio de la Frontera Sur (ECOSUR), San Cristobal de las Casas, Chiapas, México for support in developing the algorithm to delineate the boundaries of FMV. We extend our sincere thanks to the members expert committee, Department of Biotechnology for valuable inputs during the progress presentations of the research on field margin vegetation and Socio-Ecological Environment—Structural, Functional and Spatio-temporal Dynamics in Rural–Urban Interface. We wholeheartedly thank all the field and administrative staffs for their support, valuable inputs, immense help and cooperation during the entire period of study.

The authors expect that this book will be quite useful to the stakeholders involved in research and development of agricultural landscapes in India for formulating policies for sustainable socio-ecological development. This will substantially contribute to the knowledge base among scientific community worldwide and encourage further study in this field of research.

We would like to extend our thanks and appreciations to Project Coordinator, Doerthe Mennecke-Buehler, Chandra Sekaran Arjunan, Production Editor, Scientific Publishing Services, for the kind help, support, and assistance during the entire publication process and series editors—Ulrich Förstner, Wim H. Rulkens, and Wim Salomons for their careful reading of the book manuscript. We extend our sincere thanks to Ms. S. Mano Priya, Project Manager, Scientific Publishing Services, India, and the production team for the cooperation and efficient support during the production of this volume.

Bengaluru, India   Prof. Sunil Nautiyal
                                                              Dr. Mrinalini Goswami
                                                              Puneeth Shivakumar

Contents

1   **Introduction to Field Margin Vegetation (FMV)** 1
     1.1   Agroecosystems and Biodiversity 1
     1.2   Field Margin Vegetation 2
     1.3   Research on FMV 4
     1.4   Importance of FMV 5
     1 5   Significance in Indian Socioecological Systems 8
     1.6   Types of FMV 9
     References 12

2   **Urbanization and Peri-Urbanization in Bengaluru** 17
     2.1   Pattern of Urbanization in India 17
     2.2   Urban Primacy; The Case of Bengaluru, Karnataka 18
     2.3   The Push–Pull Paradigm 20
     2.4   Rural-Urban Interface and Concept of Peri-Urbanization 23
     2.5   Fate of India's Rural-Urban Interfaces 24
     2.6   Bengaluru's Experience in Rural-Urban Interface 25
     2.7   Looking Forward 26
     2.8   Identifying Agroecosystems in Rural-Urban Interface 27
     2.9   Climate, Forest and Physiography 29
     2.10   Land Use and Agriculture 31
     2.11   Demography and Socioeconomics 36
     References 37

3   **Agroecosystems in Rural-Urban Interface** 41
     3.1   Trade-offs in Peri-Urban Agroecosystems 41
     3.2   Socioecological Household Survey and Assessment of FMV 43
     3.3   The Changing Field Margins in Rural-Urban Interface
           of Bengaluru 44
     3.4   Socioeconomic Landscape and Direct Economic Benefits
           from FMV 49
     References 56

4   Structure and Functions of FMV in Rural–Urban Interface   57
    4.1   Phytosociological Study of Field Margins   57
    4.2   FMV Species Richness and Diversity Indices   59
    4.3   Manipulation and Management of FMV   69
    4.4   Traditional Ecological Knowledge on Field Margin
           Vegetation   71
    4.5   Way Forward for Enhancing FMV Value and Preserving TEK   ...   74
    References   75

5   Spatio-Temporal Dynamics of Rural-Urban Interface and FMV   77
    5.1   Land Use Land Cover Analysis of Rural-Urban Interface
           of Bengaluru   77
    5.2   Methodology and Data Used   78
    5.3   LULCC in Rural-Uurban Interface of Bengaluru Over Four
           Decades   79
    5.4   Interclass Decadal Change of LULC   81
    5.5   Village-Wise LULC Change   84
    5.6   Vegetation Indices; NDVI and SAVI   89
    References   94

6   Delineation and Monitoring of FMV   95
    6.1   Remote Sensing in Vegetation Study; Prospect for FMV
           Mapping   95
    6.2   Methodology for Delineating Field Margin Vegetation   98
           6.2.1   Post Processing   106
    6.3   Delineation of FMV and Accuracy Assessment   106
    References   111

7   Overview of a Few Important FMV Species and Crop
    Influencing FMVs of Rural–Urban Interface of Bengaluru   115
    7.1   *Pongamia pinnata*   115
    7.2   *Eucalyptus* spp. in Field Boundaries   116
    7.3   *Vitis* spp.   119
    References   120

8   Strategizing FMV Conservation for Sustainable
    Agroecosystems in Rural-Urban Interface   121
    8.1   Why to Retain FMVs and Traditional Agroecosystems   121
    8.2   Strategies for Developing Sustainable Agroecosystems
           Through Conserving FMV   124
    8.3   Way Forward   126
    References   127

Annexure A: Tree Species in Six Villages of the Northern Transect
of Bengaluru   129

Annexure B: Shrub Species in Six Villages of the Northern
Transect of Bengaluru 135

Annexure C: Herb Species Listed in Six Villages of the Northern
Transect of Bengaluru 139

Annexure D: Percentage of Farms Having Tree in the Field
Margins for Different Crop Groups and Presence of FMV Tree
Species (From the Selected Fields of the Northern Transect
of Bengaluru) 145

Annexure E: Relative Density of Herb Species in Six Study
Villages of the Northern Transect of Bengaluru 149

Annexure F: Relative Frequency of FMV-herb Species in Six
Study Villages of the Northern Transect of Bengaluru 153

Annexure G: Photo Plates-Some FMV Species Prevalent
in the Northern Transect of Bengaluru 157

Annexure H: Correlation Between FMV Products
and their Economic Benefits in the Northern Transect
of Bengaluru 163

Bibliography 167

List of Figures

Fig. 1.1   General illustration of field margin 3

Fig. 2.1   Land use land cover maps of three major urbanized districts of Karnataka. *Source* Indian Earth Observation Visualization 22

Fig. 2.2   Location of study area in northern Bengaluru, India 28

Fig. 2.3   Decadal Change in crop area under cereals and pulses in Bengaluru Rural and Urban districts 32

Fig. 2.4   Area under fruit and vegetable crops in Bengaluru Rural and Bengaluru Urban districts 33

Fig. 2.5   Area under oil crops in Bengaluru Rural and Bengaluru Urban districts 33

Fig. 3.1   Respondents on FMV area engulfed for agricultural expansion. *Source* Authors' estimation 45

Fig. 3.2   Status of change in agricultural land among the farming households. *Source* Authors' estimation 46

Fig. 3.3   Percentage of households stating different reasons for change in agricultural landscape. *Source* Authors' estimation 47

Fig. 3.4   Distribution of households under different income classes in the study transect 50

Fig. 3.5   Average landholding per household in the study transect 50

Fig. 3.6   Economic benefit from FMV in terms of foods in Rs per household (HH) per year. *Source* Field survey 52

Fig. 3.7   Use of FMV-based resources in peri-urban landscape of Bengaluru, India. *Source* Field survey 53

Fig. 3.8   Use of FMV-based resources (estimated raw weight) in peri-urban landscape of Bengaluru, India. *Source* Field survey 53

Fig. 4.1   Methodological framework for studying rural–urban interface of Bengaluru describing phytosociological assessment, socioecological survey and land use land cover analysis 58

Fig. 4.2   Number of species under different plant types
            across the study transect 60
Fig. 4.3   **a–f** Density of tree species per hectare in six villages
            of the northern transect of Bengaluru 64
Fig. 4.4   Number of plant species from used to acquire various
            economic benefits in the northern transect of Bengaluru 73
Fig. 4.5   Use of FMV species for medicinal purposes in the northern
            transect of Bengaluru 73
Fig. 5.1   Methodological framework for decadal land use land cover
            change estimation in the northern transect of Bengaluru 78
Fig. 5.2   Percentage of area under different land use land cover
            classes at four points of time in the northern transect
            of Bengaluru 80
Fig. 5.3   Decadal land use land cover change in the selected transect
            of northern Bengaluru 81
Fig. 5.4   **a** Land use and land cover of Konaghatta, **b** Land use
            and land cover of Heggadihalli, **c** Land use and land cover
            of Kundana, **d** Land use and land cover of Arakere (Arekere),
            e Land use and land cover of Muddenahalli, **f** Land use
            and land cover of Singanayakanahalli 86
Fig. 5.5   Share of FMV areas to crop areas across the transect (based
            on the analysis of the sampling plots) 89
Fig. 5.6   **a** Results of manual digitization of FMV for 2005, **b** Results
            of manual digitization of FMV for 2007 90
Fig. 5.7   **a** and **b** Comparison of NDVI versus SAVI at 30 m spatial
            resolution for northern transect of Bengaluru 92
Fig. 6.1   Steps involved in developing accurate methodology
            for delineating field margin vegetation 99
Fig. 6.2   Location and satellite image of Pilot study site located
            in northern Bengaluru Location and satellite image of pilot
            study site located in northern Bengaluru 101
Fig. 6.3   **a** Input for algorithm 1 and 2 (panchromatic image). **b** Input
            for algorithm 3 (classified image) 102
Fig. 6.4   **a** Histogram for classified multispectral image. **b** Histogram
            for panchromatic image 105
Fig. 6.5   Result of supervised maximum likelihood classification
            (MLC) for algorithm 3 107
Fig. 6.6   Result of manual digitization of FMV 108
Fig. 6.7   Output images for step 2 and 3 of the methodological
            framework for all the three algorithms 109
Fig. 7.1   Use of *Pongamia* parts for various purposes in different
            zones of northern transect of Bengaluru 116
Fig. 7.2   Variables describing the relative dominance of *Eucalyptus*
            spp. in three zones of northern transect of Bengaluru 117

Fig. 7.3   Presence of FMV tree per plot of grape field versus all crop
            fields 120

List of Tables

Table 1.1   Literature review matrix showing the availability
of research in different context 4

Table 2.1   Urbanization trend in Karnataka 18

Table 2.2   Pattern of urbanization in selected districts of Karnataka 19

Table 2.3   Economic performance of selected districts of Karnataka 19

Table 2.4   Land use land cover change in one decade in three districts 21

Table 2.5   Monthly rainfall (mm) and temperature (°C) pattern
of Bengaluru Rural and Bengaluru Urban districts
(100 years' average) 30

Table 2.6   Pattern of change in area (in Ha) under different crops
in Bengaluru Urban and Bengaluru Rural districts 32

Table 2.7   Description of socioeconomic parameters of selected
villages 36

Table 3.1   Details of study area for the field margin vegetation
in the northern transect of Bengaluru 43

Table 3.2   Percentage of plots under different crops 45

Table 3.3   Change in area under different crops in three zones
of the transect 48

Table 3.4   Number of FMV trees present in agricultural plots
of different crops 48

Table 3.5   Percentage change in area under different land use land
cover classes from 1991 to 2018 49

Table 4.1   Family-wise distribution of number of species
documented in each village 61

Table 4.2   Relative density (%) of shrub species across the northern
transect of Bengaluru 65

Table 4.3   Relative density and relative frequency of FMV tree
species present in six villages of northern transect
of Bengaluru 67

Table 4.4   Diversity and evenness of FMV tree species in six
villages of northern transect of Bengaluru 69

Table 4.5   Basal area and stem volume of FMV species in transition
            zone of northern transect of Bengaluru   70
Table 4.6   Types of vegetation on field boundaries in different zones
            in the northern transect of Bengaluru   70
Table 4.7   Percentage of different types of FMV present in the three
            sections of the northern transect (n = 60), Bengaluru   71
Table 5.1   Inter-class transition matrix of LULC (1991–2001)
            in the northern transect of Bengaluru (area in percentage)   82
Table 5.2   Inter-class transition matrix of LULC (2001–2011)
            in the northern transect of Bengaluru (area in percentage)   82
Table 5.3   Inter-class transition matrix of LULC (2011–2018)
            in the study transect (area in percentage)   82
Table 5.4   Percentage of area under different land use and land cover
            classes of six villages located in three different zones
            of northern transect, Bengaluru   84
Table 5.5   Details of satellite data used for vegetation
            assessment in the northern transect of Bengaluru   90
Table 5.6   Analysis of vegetation cover analyzed using NDVI
            and SAVI for the northern transect of Bengaluru   94
Table 6.1   Description of data used in the study of northern transect,
            Bengaluru   98
Table 6.2   Data and parametric description of three algorithms
            developed to delineate the FMV in northern transect
            of Bengaluru   104
Table 6.3   Results of accuracy assessment for all the three algorithms   110
Table 6.4   Accuracy assessment of FMV for three different
            algorithms   110
Table 7.1   Density and frequency of *Pongamia pinnata*
            across the transect   116
Table 8.1   Understanding on field margin vegetation in rural-urban
            interface of Bengaluru, India   123
Table A.1   Tree species listed in six villages of the northern transect
            of Bengaluru   130
Table B.1   Shrub species listed in six villages of the northern transect
            of Bengaluru   136
Table C.1   Herb species listed in six villages of the northern transect
            of Bengaluru   140
Table D.1   Percentage of farms having tree in the field margins
            for different crop groups and presence of FMV tree
            species in the northern transect of Bengaluru   146

Chapter 1
Introduction to Field Margin Vegetation (FMV)

1.1 Agroecosystems and Biodiversity

Agroecosystems are managed ecosystems where human has intentionally and selectively altered the natural composition of living organisms for food, fodder, fiber or other economic benefits. Thus, such ecosystems have significant social, economic and ecological dimensions. It is an interface between human society and natural ecosystem. While integrating the various components of an agricultural landscape, the interacting elements, viz.- croplands, livestock, grazing land, neighboring forest and households, agroecosystem can be considered as a single functional unit. Compared to natural ecosystems, agroecosystems have high fluidity, vulnerability, spatiotemporal differences, poor stability and low biodiversity (Zhu et al. 2012). Although it is an integral part of larger ecosystem, it has low resilience compared to the natural part of the system. Ecosystems with higher diversity are more stable because they exhibit higher resistance and resilience which help them to withstand undesirable changes or provide ability to recover following disturbance (Nicholls and Alteiri 2004). Therefore, agroecosystem requires deliberate conservation effort to maintain the function and services pertaining to ecological and human sub-systems.

Agricultural landscapes are significant in terms of biodiversity, food security and livelihoods. Agricultural landscapes support wild biodiversity by providing food and shelter, have importance in socioecological system including human culture, spirituality and recreation. Biodiversity of agricultural landscape consisted of species and genetic diversity from both agricultural and non-agricultural land and water. The richness in biodiversity in agroecosystems is contributed by multi-element structure and heterogeneity (Marshall and Moonen 2002). Non-crop agro-biodiversity include all non-domesticated plant and animal species within the crop land and in the surrounding area which is part of the agro-ecosystem. The four principal components of agrobiodiversity are genetic resources, biodiversity that supports the ecosystem services in agroecosystems (agricultural, pastoral, forest and aquatic), abiotic factors and socioeconomic dimensions (Zimmerer 2014). The beneficial agrobiodiversity may include a tree providing shades to earthworms and microorganisms increasing soil fertility to pollinators in the shrubs at the field margin. In India, agricultural landscapes are classified into agroecological zones, which are curved out of climate zones, and further 20 regions are identified based on physiographic features, soil characteristics, bio-climatic features and length of growing periods. National Bureau of Soil Survey and Land Use planning (ICAR) have sub-divided the regions into 60 agroecological sub-regions in India. The physiographic and climatic heterogeneity, cultural diversity and agricultural practices have contributed towards a diverse agrobiodiversity country. These highly diverse ecosystems provide conditions for the adaptation of a large number of both indigenous and exotic crop plants bringing a revolution in the growth of agriculture and agrobiodiversity (Singh 2017). Landscape change, change in consumer behaviour, climate change, market-oriented cultivation etc., are some of the determinants of agrobiodiversity in landscapes with increasing human intervention. With the change in types and pattern of domesticated biodiversity for improving economic benefits, the biodiversity in the surroundings also deviates from the original through natural selection.

1.2   Field Margin Vegetation

Field edges consist of a variety of sub-habitats such as hedgerows, wooded areas, stream edges and ditches. Field margins increases the local biodiversity in terms of both floral species and invertebrates. The structure and composition of field margin vegetation are determined by a wide range of land use intensity and landscape parameters (Billeter et al. 2008). These may include cropping pattern, nutrient inputs, types and number of livestock, farm size, intensity of agrochemical use, distance to other neighbouring land use, proximity to natural vegetation, area of semi-natural patches etc. The influencing factors and structure of field margins are diverse and thus the definitions are also varied (Marshall and Moonen 2002).

Field Margin Vegetation (FMV) such as hedgegrow, buffer strips, modified wood lots etc., contain the most ecologically significant part of non-crop biodiversity in agricultural landscape. Despite of small size of field boundaries as compared to crop area, they show a great richness of organisms. The vegetation on the field margins are affected by many biotic and abiotic factors; interaction of ecological and economic elements, landscape structure, cropping practices and farm types are some of them. The type and quantity of vegetation at the boundary have also an impact of on the crops. Field Margins are uncultivated areas of natural or semi-natural vegetation located adjacent to the field. They are the interface between the agriculture and another other land use land cover type. The field margins form an integral part of the non-cropped habitats and harbour variety of floral and faunal species (Marshall 1987; Rands 2011). General illustration of field margin is presented in Fig. 1.1.

The field margins are the pre-existing boundary of a cropland which can be termed in different ways depending on their structure, function and conservation approach.

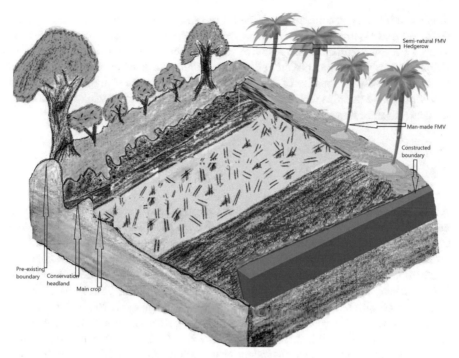

Fig. 1.1   General illustration of field margin

The diversity of conservation management approaches for field margins can be best summarised as follows (Greaves and Marshall 1987):

(a)   Boundary strips
(b)   Grass and wild flower strip
(c)   Sterile strip
(d)   Uncropped wildlife strip
(e)   Set-aside margin
(f)   Mixture of sown and wildlife
(g)   Crop edge.

Vegetative strips are area with vegetation set aside from main crop area within or around a field and have multiple benefits on biota, habitat improvement, shade tree, carbon stock, yield, biomass production, environmental quality and socioeconomics of farmers (Haddaway et al. 2016). These are primarily man-made interventions for ecological and economic benefits. With intensification of agriculture and changing agricultural practices, there have been loss of biodiversity and consequent decline in ecological services provides by those flora and fauna. Increase in agricultural chemicals (herbicides, fertilizers, pesticides etc.) and clearing of non-crop vegetation as required for new crops also contribute towards diminishing diversity of agrobiodiversity. Sown grass and wildflower strips are efforts to enhance the ecosystem services of agrobiodiversity, specifically certain groups of species of birds, insects and plants (Haaland and Gyllin 2011). Different types and structure of field margins can be utilised for environmental and resource enhancement including establishing grass strips, sowing wildflower pollen and nectar sources for insects or sources of seed for birds. Multi-functionality of landscape can be enhanced through managing and developing field margins with endemic and traditionally maintained plants useful for farmers and the ecosystem.

1.3 Research on FMV

A few researches on management of field margins has been extensively carried out in European countries with an aim of field level biodiversity conservation and enhancing agronomic benefits. The aspects of literatures reviewed on the topics are presented in Table 1.1. Research on field margin vegetation with regard to structure and composition (Kleijn and Snoeijing 1997; Kleijn and Verbeek 2000; Aude et al. 2003; Helenius and Backman 2004; De-Cauwer et al. 2006; Tansey et al. 2009; Marshall and Moonen 2002; Fridley et al. 2009; Noordijk et al. 2011; Balzan et al. 2016; Alignier and Aviron 2017; Alignier 2018), interaction with agriculture (Fried et al. 2018; Bischoff et al. 2016; Alignier and Baudry 2015; Balzan and Moonen 2014; De-Cauwer et al. 2006; Marshall et al. 2006; Marshall 2004; Marshall and Moonen 2002; Holland and Luff 2000; Kleijn and Snoeijing 1997), factors affecting field margin vegetation (Fritch et al. 2011; Tarmi et al. 2009; Aavik et al. 2008; Le Coeur et al. 1997), flora-fauna relationship (Balzan et al. 2016; Morelli 2013; Marshall et al. 2006; Woodcock et al. 2005; Asteraki et al. 2004; Roy et al. 2003; Perkins et al. 2002; Thies and Tscharntke 1999; Sparks and Parish 1995) and other

Table 1.1 Literature review matrix showing the availability of research in different context

Research component on FMV	European countries	Other countries	India
Structure & composition	✔	✔	Studies are broadly on agroforestry which do not directly deal with FMV as separate entity
Species and habitat	✔	–	
Flora-fauna relation	✔	–	
Impact on agriculture	✔	✔	
Economic benefit	–	–	
Other ecosystem services	✔	–	
Driving forces	✔	–	
Impact of agricultural practices	✔	–	
Conservation	–	–	

ecosystem services (Hackett and Lawrence 2014; De-Cauwer et al. 2006; Douglas et al. 2009; Olson and Wäckers 2007; Le Coeur et al. 2002) are available in different western countries, but not much in India. In Indian perspective it is little hard to find the studies on economic importance and valuation of field margin vegetation. Most of the studies have field sites located in European countries. Research on FMV with relation to agriculture has primarily focused on insect habitat, pollinating agents and pest control, impact of fertilizers and pesticides etc. In terms of flora-fauna relationship, forages and habitat provided by FMV have been studied.

Habitat manipulation in the field margin change the pest as well as predator diversity affecting the crops. Research suggests that biological control and pest suppression of field may be enhanced by vegetation near boundary. The influence of natural, semi-natural or manmade vegetation strips in field boundary has to be focused and need to consider the interaction between crops and FMV. The varied strategies used for managing field margins influences the benefits within the margin as well as on its surroundings. There is limited research investigating the influence of field margin to the crops. However, existing studies have shown positive effects of FMV on cropping systems. Field margins management is one of the most frequently adopted options within agri-environment schemes in European countries (Boatman 1999). Research suggests that biological control and pest suppression of field may be enhanced by vegetation near boundary. The influence of natural, semi-natural or manmade vegetation strips in field boundary has to be focused and need to consider the interaction between crops and FMV. In agricultural point of view, field margins are considered to have less economic value than the crops cultivated within fields. The management of field margin requires additional effort and monetary inputs (Snoo 1994); thereby it is neglected widely. India is lagging behind in the management strategies of field margins and extensive research has to be carried to achieve biodiversity conservation on field margins in an economical and logistical manner.

Roles of field margin vegetation and their interaction with different faunal species, specifically, Amphibia, Reptilia, Arthropods etc., have not been covered in previous studies. It is also essential to understand the functions of FMV in controlling or providing habitat for pest, prey, predators and pollinating agents in the farm. It's function on ecosystem stability is to be evaluated along with the contribution towards the socioeconomic development of the community. The effect of dynamics of complex urbanizing landscape on structure and composition is yet to be evaluated for any country. Driving forces for the change in structure and functions of FMV, such as socioeconomic conditions of the farmers, changing climatic parameter and market driven forces in ensuing agricultural shift assumed to have immense impact on field margin vegetation.

1.4 Importance of FMV

Field margins have an essential multi-functional role to protect soil and water, enhance biodiversity, and provide food, fodder, fuel and other products in agricultural landscapes. Initially, field margins were considered as source of weeds, pests and diseases (Marshall 1987). The non-crop floral diversity on the boundary have a role in providing habitats to pests and also work as a support to natural pest controlling agents. The different species in the field boundary act as hosts for pests and predators, aiding in the functions of an agroecosystem. Moreover, FMV provide sheets for crops by preventing erosion, acting as a windbreak or barrier to the non-point sources of pollution into the field, so retaining it is important. It clearly defines the field boundary and it protects the land from the depletion of nutrients and water resources. In the ancient days, agricultural lands were delineated by field margins to represent the land ownerships and protect from wild animals (Marshall 2004). Smith et al. (2008) mentioned three ecological functions of field margins are to increase species density in an agroecosystem (biodiversity value), provide habitats for rare or endangered species (conservation value) and enhance ecosystem services such as pest control and decomposition (functional value). The importance of vegetation on field boundary as a reservoir for pests and predators has been widely accepted. It helps to enhance the richness of biodiversity to increase resistance and resilience of the ecosystem. Many studies have demonstrated increased abundance of natural enemies and more effective biological control where crops are bordered by wild vegetation (Altieri 1999). Among the multitude of functions of FMV, some are traditionally known to the farmers, for which they want to retain the vegetation on field boundary. Some have been identified and assessed through ecological research, may not be known to the farmers; although, the farmers and the ecosystem as a whole have been receiving indirect benefits.

Continuous arable farming of annual crops is least beneficial and sometimes harmful to the species diversity in margins and careless usage of fertilizers and herbicides may be one of the reasons for the depletion of area under FMV. Moreover, farmers may clear existing vegetation cover with margins prior to the planting of crops for economic benefits. Continuous removal of vegetation negatively impacts landscape and efficiency of agriculture. Naturally segmented field margins within study area were characterized by relatively heterogeneous plant communities. International studies suggest that presence of weedy filed margin may be more effective initially at enhancing biological control and pest suppression in comparison with annual flowering field margins (Balzan and Moonen 2013). Weed species play a vital role for supporting cropping systems. The previous studies suggest that compared to conventional farming, traditional farming results higher species diversity in the field margins. According to the studies conducted in Finland found that minimum width of field margin should be 5 m and it effectively improves the biodiversity value.

FMV limits the effects of water and wind erosion and acts as a buffer, reduces non-point pollutions and creates refuges for many species of fauna and flora that are not part of crop diversity in the neighbouring agricultural land. Apart from delineating the field limit, field margin vegetation has secondary roles in terms of agronomical, environmental, conservation and recreational interests (Marshall 2004). Field margins serve as shelter; provide food, fodder, fuelwood and medicine; and support other ecosystem services necessary for agriculture (Nautiyal et al. 2020). Moreover, field margins promote ecological stability by exploiting the pest predators and para-sites (Marshall 2004; Wilson 1995). Some of the traditionally known functions and knowledge about maintaining FMV are:

(a)   Defines farm boundary
(b)   Prevents trespassers, wild animals and livestock attack
(c)   Provide food, fodder, fibre, mulch and medicines
(d)   Shades and windbreaks for crops
(e)   Reduce nutrient loss and soil erosion
(f)   Maintain water balance
(g)   Growing plants that are non-competitive to crops in terms of nutrients and water
(h)   Growing beneficial plants that shelter useful fauna
(i)   Not to grow plants that harbour pests and diseases.

The functions assessed or identified through scientific and ecological investigations:

(a)   Promotion of ecological stability and enhance resilience
(b)   Enhance useful organisms like pollinators
(c)   Exploit pest-predators
(d)   Reduce weed invasion and reduce use of pesticide and herbicide
(e)   Maintain agrobiodiversity
(f)   Promote biodiversity and conservation
(g)   Encourage subsidiary agri-business
(h)   Maintain multi-functionality of landscape
(i)   Maintenance of socio-culturally important species and historical features.

Agroforestry systems or other perennial plantings on agricultural lands play a significant role in storing vegetative carbon (Albrecht and Kandji 2003), so does the FMVs too. Field margins have the potential to help mitigate greenhouse gas emissions from agricultural activities through carbon (C) sequestration in the woody biomass of trees and shrubs as well as in the soil (Thiel et al. 2015). India is a vast country with a rich biological diversity where agriculture is the land use with highest area. India has approximately 159–160 Million Ha of agricultural land, out of which, almost 110 Million Ha is historically supported directly or indirectly by traditional socio-ecological system. As a part of Paris Agreement (COP-21), India is committed to achieve an additional Carbon Sink of 2.5–3 Billion Tonnes of Carbon-di-oxide Equivalent by the year 2030. 50% of that can be achieved through conservation of existing trees and farm margin vegetation. Thus, the current need of conserving farm margin vegetation is aligned with the objectives of India's climate policy.

1.5   Significance in Indian Socioecological Systems

A wide range socioeconomic parameters are associated with field margin vegetation in India. Along with providing economic and ecological benefits, FMV enhances the sense of land ownership, does boundary demarcation and provides other cultural and religious values. In Indian context the major reasons for traditionally protecting trees in farmlands can be improvement of agricultural production in traditional agroecosystems, subsistence products like fodder, fuelwood etc., and additional income (Saxena 1992). FMV complement the traditional farming system of India agricultural in yield management by increasing soil fertility by nutrient recycling, water conservation, mulching and providing shade for crops and farmers/farm workers. Farmers maximise their income by multi-storied cropping such as inter-cropping of perennial crops like coconut, areca nut etc., with seasonal crops like vegetables, banana etc. The non-tree species both naturally grown and planted in the field margins, have also immense socioecological significance. Those have food, medicinal and other sociocultural values, for which the farmers generally try to refrain them from removing those vegetation from field boundaries. In India, a live boundary of a farm can have plants such as shikakai (*Acacia concinna*) and mehndi (*Lawsonia inermis*), not only protects the farm from humans and animals but also provides biomass for other usage (Cole et al. 2017). The promotion of agroforestry under government schemes have altered those traditional field boundaries through introduction of economically important species like *Eucalyptus* and Silver Oak with consequent discount of social and ecological values of field boundaries (Nautiyal et al. 2020).

Cultivation and management of non-crop plants in agricultural land in the landscape is a traditional practice. Study in Uttara Kannada district (Shastri et al. 2002) shows that the species in agroecosystems are similar to the species in nearby forest; which indicates that the farmers have developed the agroecosystems as a mimic of natural ecosystems. It is characterised by an intensive integration of variety of plants to ensure sustained availability of multiple products as direct benefits such as food, vegetables, fruits, fodder, fuel, foliage, medicine, and raw materials for agricultural implements. Their findings also show that farmers have the tendency to retain multifunctional tree species in agricultural fields. The types of trees in farms and agroecosystems are product of interaction between local and formal knowledge acquired by the farmers and they domesticate the species necessary for survival and subsistence (Pandey 2002). Study carried out in a cotton farm of Telangana (Flachs 2015) revealed nearly 100 semi-managed species of vegetables, trees and wild plants; where, each household manages an average of 17 of non-crops species in their farms for economic benefits through various uses. In-spite of commodification and increased monoculture of agriculture, farmers continue to preserve the non-crop diversity in their agricultural field.

Although significance of non-crop plants in agroecosystem is immense, it has not been studied adequately in India as a separate entity as it has been investigated in Europe and North America (Zuria and Gates 2013; Wuczyński et al. 2014; Guiller et al. 2016; Balzan et al. 2016). To acknowledge the significant role of FMV in the socioecological system and its contribution towards climate change mitigation and adaptation, it is essential to map and assess FMV. The importance of traditional ecological knowledge (TEK) is well established within natural resource management and sustainability science (Zent and Maffi 2009). A combination of scientific knowledge and TEK can play a positive role in sustaining biodiversity and nature's services and subsequently sustain the cultural norms of a community (Becker and Ghimire 2003). Traditional societies have a holistic view of the ecosystem and the social system where relationship with nature is based on coexistence (Ramakrishnan 2005). Thus, the traditional systems help the communities to develop adaptive strategies (Berkes 2004) to cope with any change in the biotic or abiotic environment and sustainable use of resources.

The unaccounted environmental benefits of field margin vegetation in India are wide-ranging. We can state one instance from our observations regarding crop residue burning. Uttar Pradesh is the largest producer of residues from cereal crops (72.02 MT/year) followed by Punjab with 45.58 MT per year (Jain et al. 2014). The dimension of effects of cereal residue burning in Punjab and Haryana is much bigger as compared to other states. In Punjab and Haryana, 80% of rice straw was burnt in situ followed by Karnataka (50%) and Uttar Pradesh (25%) (Gupta et al. 2003). Mechanized harvesting has contributed largely to this hazard where, removal of vegetation from the boundary happened during agricultural intensification. It can be related that absence of trees at the field margins in Punjab and Haryana have made the in-situ residue burning an easier process with huge environmental effects. Trees on the margin indirectly prevent residue burning; for ethical, economic and ecological values of trees that is well-known to farmers, don't allow them to burn residue in the farm fields. Therefore, an insight to these values of field margins which are likely to get changed with changing agricultural landscape should be developed comprehensively to formulate strategy for conservation of this significant component of biodiversity in agroecosystems.

1.6   Types of FMV

Field margins can be vegetated and non-vegetated. Various factors including types of use, pattern of adjacent land uses, use of agro-chemicals, availability of nutrients and water, kind of protection required etc., play role in determining the pattern of field boundaries. Non-vegetated boundaries may be a concrete pathway, wooden, bamboo, pillar or wire fencing or a concrete wall. Dead hedges are used in the boundary to give protective barriers to the crops. Living hedges comprising of planted shrubs and naturally grown herbaceous plants are also present in the study area. *Agave* spp. are found planted in a few boundaries which act as a very efficient protection for the crops and give economic benefits in the form of fibre as well. Our observation indicated that in some cases waterbodies (like pond), banks and edge of woodlot act as a boundary for crop fields hence such structural attributes in landscape do not

Photo 1.1 Natural FMV
with dominance of weeds

provide adequate space for distinct FMVs vegetated field margins prevalent in the study area may have the following pattern of plants or combination of those.

Natural field margins *Natural vegetation refers to a plant community which has grown naturally without human aid and has been left undisturbed by humans for a long time. And it is noted that in the field margin vegetation are not irreversibly affected by major human interventions, such as the introduction of exotic species or soil removal. One typical natural farm boundary can be seen as in Photo 1.1.*

Semi-natural vegetation *in agricultural land mainly includes extensively managed grasslands, agro-forestry areas and all vegetated features that are not used for crop production. Hence semi-natural vegetation includes vegetation due to human influences but which has recovered to such an extent that species composition and environmental and ecological processes are indistinguishable from, or in a process of achieving, its undisturbed state. Combination of economically important plants like coconut with naturally grown understory shrubs and herbs are also evident in the villages. Fodder grass is another category of flora planted for their economic benefit and associated advantage of barrier. (Photo 1.2)*

Photo 1.2   Semi-natural
FMV

Photo 1.3   Planted field
margins FMV-a

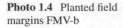

Photo 1.4 Planted field
margins FMV-b

Artificial (planted) field margin vegetation are those boundary areas of crop fields which are completely modified by human. Such margins are only utilized by plantation of perennial crops or short duration crops for economic benefits. These are the vegetation planted in the field margins for timber, fuelwood or for any other economic purpose. Photos 1.3 and 1.4 are representatives of planted field margins of city peripheral areas of northern Bengaluru Urban District.

References

Aavik T, Augenstein I, Bailey D, Herzog F, Zobel M, Liira J (2008) What is the role of local landscape structure in the vegetation composition of field boundaries? Appl Veg Sci 11(3):375–86

Albrecht A, Kandji ST (2003) Carbon sequestration in tropical agroforestry systems. Agric Ecosyst Environ 99(1–3):15–27

Alignier A (2018) Two decades of change in a field margin vegetation metacommunity as a result of field margin structure and management practice changes. Agric Ecosyst & Environ 251:1–0

Alignier A, Aviron S (2017) Time-lagged response of carabid species richness and composition to past management practices and landscape context of semi-natural field margins. J Environ Manag 204:282–90

Alignier A, Baudry J (2015) Changes in management practices over time explain most variation in vegetation of field margins in Brittany France. Agric Ecosyst Environ 211:164–172

Altieri MA (1999) The ecological role of biodiversity in agroecosystems. In: Invertebrate biodiversity as bioindicators of sustainable landscapes. Elsevier, pp 19–31

Asteraki EJ, Hart BJ, Ings TC, Manley WJ (2004) Factors influencing the plant and invertebrate diversity of arable field margins. Agric Ecosyst Environ 102(2):219–231

Aude E, Tybirk K, Pedersen MB (2003) Vegetation diversity of conventional and organic hedgerows in Denmark. Agric Ecosyst Environ 99(1–3):135–147

Balzan MV, Moonen AC (2013) Field margin vegetation enhances biological control and crop damage suppression from multiple pests in organic tomato fields. Entomol Exp Appl 150:45–65

Balzan MV, Bocci G, Moonen AC (2016) Landscape complexity and field margin vegetation diversity enhance natural enemies and reduce herbivory by Lepidoptera pests on tomato crop. BioControl 61(2):141–54

Becker CD, Ghimire K (2003) Synergy between traditional ecological knowledge and conservation science supports forest preservation in Ecuador. Glob Ecol Conserv 8(1):1

Berkes F, (2004) Rethinking Community-Based Conservation. Conserv Biol 18, 621–630. https://doi.org/10.1111/j.1523-1739.2004.00077.x

Billeter R, Liira, J, Bailey D, Bugter R, Arens P, Augenstein I, Cerny M (2008) Indicators for biodiversity in agricultural landscapes: a pan-European study. J Appl Ecol 45(1):141–150

Bischoff A, Pollier A, Lamarre E, Salvadori O, Cortesero AM, Le-Ralec A, Tricault Y, Jaloux B (2016) Effects of spontaneous field margin vegetation and surrounding landscape on Brassica oleracea crop herbivory. Agric Ecosyst Environ 223:135–143

Boatman ND (1999) Field margins and buffer zones: ecology, management and policy. Ann Appl Bio l54:345–352

Cole LJ, Brocklehurst S, Robertson D, Harrison W, McCracken DI (2017) Exploring the interactions between resource availability and the utilisation of seminatural habitats by insect pollinators in an intensive agricultural landscape. Agric Ecosyst Environ 246:157–67

De-Cauwer B, Reheul D, Nijs I, Milbau A (2006) Dry matter yield and herbage quality of field margin vegetation as a function of vegetation development and management regime. J Life Sci 54(1):37–60

Douglas DJ, Vickery JA, Benton TG (2009) Improving the value of field margins as foraging habitat for farmland birds. J Appl Ecol 46(2):353–62

Flachs A (2015) Persistent agrobiodiversity on genetically modified cotton farms in Telangana, India. J Ethnobiol 35(2):406–427

Fridley JD, Senft AR, Peet RK (2009) Vegetation structure of field margins and adjacent forests in agricultural landscapes of the North Carolina Piedmont. Castanea 74(4):327–339

Fried G, Villers A, Porcher E (2018) Assessing non-intended effects of farming practices on field margin vegetation with a functional approach. Agric Ecosyst Environ 261:33–44

Fritch RA, Sheridan H, Finn JA, Kirwan L, Huallachain DO (2011) Methods of enhancing botanical diversity within field margins of intensively managed grassland: a 7-year field experiment. J Appl Ecol 48(3):551–560

Greaves MP, Marshall EJP (1987) Field margins: definitions and statistics. Field Margins 35:3–10

Guiller C, Affre L, Albert CH, Tatoni T, Dumas E (2016) How do field margins contribute to the functional connectivity of insect-pollinated plants? Landsc Ecol 31(8):1747–61

Gupta RK, Naresh RK, Hobbs PR, Jiaguo Z, Ladha JK (2003) Sustainability of post-Green Revolution agriculture: The rice–wheat cropping systems of the Indo-Gangetic Plains and China. Improving the Productivity and Sustainability of Rice-Wheat Systems: Issues and Impacts. 65:1–25

Haaland C, Gyllin M (2011) Sown wildflower strips–a strategy to enhance biodiversity and amenity in intensively used agricultural areas. The importance of biological interactions in the study of biodiversity. Rijeka: InTech 155–72

Hackett M, Lawrence A (2014) Multifunctional role of field margins in arable farming. CEA report. Cambridge Environmental Assessments

Haddaway NR, Brown C, Eggers S, Josefsson J, Kronvang B, Randall N, Uusi-Kämppä J (2016) The multifunctional roles of vegetated strips around and within agricultural fields. A systematic map protocol. Environ Evid 5(1):1–1

Helenius J, Bäckman JPC (2004) Functional diversity in agricultural field margins. Nordic Council of Ministers Copenhagen Denmark

Holland JM, Luff ML (2000) The effects of agricultural practices on Carabidae in temperate agroecosystems. Integr Pest Manag Rev 5(2):109–29

Jain N, Bhatia A, Pathak H (2014) Emission of Air Pollutants from Crop Residue Burning in India. Aerosol Air Qual Res 14:422–430. https://doi.org/10.4209/aaqr.2013.01.0031

Kleijn D, Snoeijing GIJ (1997) Field boundary vegetation and the effects of agrochemical drift: botanical change caused by low levels of herbicide and fertilizer. J Appl Ecol, 1413–1425

Kleijn D, Verbeek M (2000) Factors affecting the species composition of arable field boundary vegetation. J Appl Ecol, 37(2):256–66

Le Coeur D, Baudry J, Burel F (1997) Field margins plant assemblages: variation partitioning between local and landscape factors. Landsc Urban Plan 37(1–2):57–71

Le Coeur D, Baudry J, Burel F, Thenail C (2002) Why and how we should study field boundary biodiversity in an agrarian landscape context. Agric Ecosyst & Environ 89(1–2):23–40

Marshall EJ (1987) Field margin flora and fauna: interaction with agriculture. In: Way JM and Greig-Smith PJ (eds) Monograph No. 35. Field Margins. British Crop Protection Council, Thornton Heath, Surrey, pp 23–33

Marshall EJ (2004) Agricultural landscapes: field margin habitats and their interaction with crop production. In: Clements D, Shrestha A (eds) New dimensions in agroecology. J Crop Improv 12(1/2):365–404

Marshall EJ, West TM, Kleijn D (2006) Impacts of an agri-environment field margin prescription on the flora and fauna of arable farmland in different landscapes. Agric Ecosyst & Environ 113(1–4):36–44

Marshall EJP, Moonen AC (2002) Field margins in northern Europe: their functions and interactions with agriculture. Agric Ecosyst Environ 89(1–2):5–21

Morelli F (2013) Relative importance of marginal vegetation (shrubs, hedgerows, isolated trees) surrogate of HNV farmland for bird species distribution in Central Italy. Ecol Eng 57:261–266

Nautiyal S, Goswami M, Nidamanuri RR, Hoffmann EM, Buerkert A (2020) Structure and composition of field margin vegetation in the rural-urban interface of Bengaluru, India: a case study on an unexplored dimension of agroecosystems. Environ Monit Assess 192(8):1–6

Noordijk J, Musters CJ, van Dijk J, de Snoo GR (2011) Vegetation development in sown field margins and on adjacent ditch banks. Plant Ecol 212(1):157–67

Nicholls CI, Altieri MA (2004) Agroecological bases of ecological engineering for pest management. In: Ecological engineering for pest management: advances in habitat manipulation for arthropods. CSIRO Publishing, Collingwood, Australia, pp 33–54

Olson DM, Wäckers FL (2007) Management of field margins to maximize multiple ecological services. J Appl Ecol 44(1):13–21

Pandey DN (2002) Traditional knowledge systems for biodiversity conservation. Organization of the United Nations (FAO) Forestry Paper. FAO, Rome, Italy, pp 22–41

Perkins AJ, Whittingham MJ, Morris AJ, Bradbury RB (2002) Use of field margins by foraging yellowhammers Emberiza citrinella. Agric Ecosyst Environ 93(1–3):413–420

Ramakrishnan PS (2005) Mountain biodiversity, land use dynamics and traditional ecological knowledge. In: Huber UM, Bugmann HKM, Reasoner MA (eds) Global change and mountain regions. Adv Glob Chang Res 23. Springer, Dordrecht

Rands SA (2011) Field margins, foraging distances and their impacts on nesting pollinator success. PLoS ONE 6

Roy DB, Bohan DA, Haughton AJ, Hill MO, Osborne JL, Clark SJ, Perry JN, Rothery P, Scott RJ, Brooks DR (2003) Invertebrates and vegetation of field margins adjacent to crops subject to contrasting herbicide regimes in the farm scale evaluations of genetically modified herbicide–tolerant crops. Philos Trans R Soc Lond B Biol Sci 358(1439):1879–1898

Saxena NC (1992) Farm forestry and land-use in India: some policy issues. Ambio, 420–425

Shastri CM, Bhat DvM, Nagaraja BC, Murali KS, Ravindranath NH (2002) Tree species diversity in a village ecosystem in Uttara Kannada district in Western Ghats, Karnataka. Curr Sci 82(9):1080–1084

Singh AK (2017) Revisiting the status of cultivated plant species agrobiodiversity in India: an overview. In: Proceedings of the Indian National Science Academy 83(1)

Smith J, Potts S, Eggleton P (2008) The value of sown grass margins for enhancing soil macrofaunal biodiversity in arable systems. Agric Ecosyst Environ 127(1–2):119–125

Snoo GR (1994) Unsprayed field margins: implications for environment, biodiversity and agricultural practice: the Dutch field margin project in the Haarlemmermeerpolder. Wageningen, Ponsen and Looijen BV

Sparks TH, Parish T (1995) Factors affecting the abundance of butterflies in field boundaries in Swavesey fens, Cambridgeshire, U.K. Biol Cons 73:221–7

Tansey K, Chambers I, Anstee A, Denniss A, Lamb A (2009) Object-oriented classification of very high resolution airborne imagery for the extraction of hedgerows and field margin cover in agricultural areas. Appl Geogr 29(2):145–157

Tarmi S, Helenius J, Hyvönen T (2009) Importance of edaphic, spatial and management factors for plant communities of field boundaries. Agric Ecosyst Environ 131(3–4):201–206

Thiel B, Smukler SM, Krzic M, Gergel S, Terpsma C (2015) Using hedgerow biodiversity to enhance the carbon storage of farmland in the Fraser River delta of British Columbia. Int Soil Water Conserv Res 70(4):247–256

Thies C, Tscharntke T (1999) Landscape structure and biological control in agroecosystems. Science 285(5429):893–895

Wilson NJ (1995) The distribution of dicotyledonous arable weeds in relation to distance from the field edge. J Appl Ecol 32(2):295–310

Woodcock BA, Westbury DB, Potts SG, Harris SJ, Brown VK (2005) Establishing field margins to promote beetle conservation in arable farms. Agric Ecosyst & Environ 107(2–3):255–66

Wuczyński A, Dajdok Z, Wierzcholska S, Kujawa K (2014) Applying red lists to the evaluation of agricultural habitat: regular occurrence of threatened birds, vascular plants, and bryophytes in field margins of Poland. Biodivers Conserv 23(4):999–1017

Zent S, Maffi L (2009) Final Report on Indicator No. 2: Methodology for developing a vitality index of traditional environmental knowledge (VITEK) for the project 'Global Indicators of the Status and Trends of Linguistic Diversity and Traditional Knowledge, Terralingua'

Zhu W, Wang S, Caldwell CD (2012) Pathways of assessing agroecosystem health and agroecosystem management. Acta Ecol Sin 32:9–17

Zimmerer KS (2014) Conserving agrobiodiversity amid global change, migration, and nontraditional livelihood networks: the dynamic uses of cultural landscape knowledge. Ecol Soc 19(2)

Zuria I, Gates JE (2013) Community composition, species richness, and abundance of birds in field margins of central Mexico: local and landscape-scale effects. Agrofor Syst 87(2):377–93

Chapter 2
Urbanization and Peri-Urbanization in Bengaluru

2.1 Pattern of Urbanization in India

Among the most significant anthropogenic changes, urbanization occurs when large numbers of people start living permanently concentrated in relatively small areas resulting in urban areas, which are characterised by high level of economic activities, predominantly of non-agricultural nature. Urbanization acts as an important driver of change; both biotic and abiotic properties of ecosystems are affected due to urbanization showing it's impacts in the urban area as well as in the surroundings and also in areas far from the city (Grimm 2008). Urbanization has been a historical and spontaneous phenomenon resulted from economic activities for wellbeing of the society with consequent adverse impacts. It can be defined as a process of intensification of human settlement and their activity that results in evolution of spatial settlement system associated with the development of a socio-cultural system.

India's annual growth of urban population is 2.34% (2017–18), whereas the growth of rural population is 0.39% (2017–18). India is still predominantly rural, 65% of its population living in villages and more than 60% people depend on agriculture. The urban population of the world is 55% and is expected to increase by 13% by the year 2050. where India is projected to accommodate 416 million more people in urban areas (Desa 2018). India's steady growth of urban population from 17% in 1961 to 31% in 2011 can be attributed to both formation of new urban areas and growth of existing urban areas that are actuated demographically by rural-urban migration and spatially by inclusion of rural hinterlands (Kundu 2006). The proportion of rural population declined from 72.19% in 2001 to 68.84% in 2011 (Chandramouli 2011). The existing pace of transformation has made the urban areas to encounter constant pressure on their resources, more specifically land resources.

The impact of urbanization producing unpropitious environmental quality has been emerged as a major issue in the last three decades (Maiti and Agrawal 2005). The jeopardized environment and degrading resources are results of growth of modern cities in a haphazard way without proper urban planning and reckless industrialization (Jaysawal and Saha 2014). India's urban development lacks homogeneousness among the states where half of the urban population is concentrated in six states and high level of urbanization has been noticed in recent years in economically backward states too (Kundu 2011). With the increasing urbanization, concerns for peri-urban interface have emerged widely. Such areas are characterized by higher residential, commercial and industrial density than their bordering rural and urban areas, as well as higher rates of population growth and more dynamic processes of land conversion. In India, peri-urban areas are largely neglected in policy and practice (Randhawa and Marshall 2014), for the particular reason of these areas being at the borders, ignored as a specific area in the study of urbanization (Shaw and Das 2018), also neglected by both urban and rural administration as the delineation of borders of urban areas are blurred when it comes to population movement and resource use from rural areas. Urban planning and Master Plans or Comprehensive City Development Plan play a crucial role in sustainable urban development and urban expansion. However the planning documents legitimize the city peripheries and demarcate the urban boundaries, it is deliberately left as an unregulated "fuzzy zone" (Roy 2002). Simon (2008) emphasized on the importance and widespread nature of peri-urban interface determined by different factors like land tenure system, employment availability, income and standard of living, resource availability, role of government institutions, pattern of urbanization etc. With such complex of factors playing active role in the transition zone, it is essential to assess the landscape in a holistic way.

2.2 Urban Primacy; The Case of Bengaluru, Karnataka

In terms of urbanization, Karnataka stands 13th among all Indian states and UTs with 38.57% (in 2011) of population living in urban areas; where, Bengaluru has a share of more than 37%. The rest (63%) of the urban population is accommodated in 346 urban areas (including statutory and census towns) in the state. The state has been always ahead of national average of urban population since 1961 (Table 2.1). The

Table 2.1 Urbanization trend in Karnataka

Year	Karnataka's population	Percentage of urban population in Karnataka	Percentage of urban population in India
1961	236	22.33	17.96
1971	293	24.31	19.91
1981	371	28.29	23.33
1991	448	30.91	25.71
2001	527	33.98	27.78
2011	611	38.57	31.16

Source Census of India (2011)

increase in urban population in the state is 31.27% during 2001–2011. The population density of the state is 319 per km^2 which is much less than the national population density of 382 per km^2. Bengaluru has experienced an increase of its population from 5.69 million in 2001 to 8.50 million in 2011, with current population 12.76 million.

Disparity in urbanization across the state is very evident (Table 2.2), where Bengaluru, Dharwad and Dakshina Kannada and Mysuru districts top the list with 41–91% urban people; whereas, Kodagu, Koppal, Mandya and Chamrajanagar districts have 15–17% of their population living in urban areas. The similar pattern of disparity in economic development is also reflected among the districts (Table 2.3). Among the five divisions of districts in the state, Belgaum and Bengaluru divisions are able to attract more attention of the state for economic development resulting in higher rates of urbanization. Eswar and Roy (2018) calculated the urban primacy index for four cities and ten cities in Karnataka shows that Bengaluru urban agglomeration has more than 3.33 times urban population than that of the combined population of the next three large cities (Mysuru, Hubli-Dharwad and Mangalore), and 3.03 times more than that of next 10 large cities. The increasing urban primacy is evident where the same study shows that primacy index for four- city has increased from 2.39 in 1991 to 3.33 in 2011.

It was estimated that Bengaluru will reach a total population of 7,74 million by the end of 2015 (Taubenböck et al. 2009); but the city has shown unpredictable growth by

Table 2.2 Pattern of urbanization in selected districts of Karnataka

	Top four		Bottom four	
S. no.	District	% of urban population	District	% of urban population
1	Bengaluru	91	Kodagu	15
2	Dharwad	57	Koppal	17
3	Dakshina Kannada	48	Mandya	17
4	Mysuru	41	Chamarajanagar	17

Source Census of India (2011)

Table 2.3 Economic performance of selected districts of Karnataka

State/district	Primary		Secondary		Tertiary		GDP		Per capita
	Rs. Lakhs	% State	Rs. Lakhs	% State	Rs. Lakhs	% State	Rs Lakhs	%State	GDP (Rs)
Bengaluru	89,184	2.7	1,933,177	40.1	2,879,993	33.8	4,902,354	29.5	71,636
Belgaum	266,275	8.1	285,812	5.9	428,216	5	980,304	5.9	22,219
Dharwad	27,647	0.8	131,237	2.7	291,952	3.4	459,254	2.8	27,345
Mysuru	174,781	5.3	260,470	5.4	371,218	4.4	806,469	4.8	29,169
Karnataka	3,299,330	100	4,817,924	100	8,513,314	100	16,630,569	100	30,059

Source Directorate of Economics and Statistics, Bengaluru

having 8.5 million population in 2011. Bengaluru has joined the group of 33 megacities in the world which shelter 13% of the world's urban population. Projections show that Bengaluru may grow 2.5% annually (World Urbanization Prospects 2018). In 2011, there were 4378 people km^2, which was 2985 in 2001. Currently Bengaluru is going to be a service-oriented economy which is moving from manufacturing and industrial sectors. With secondary and tertiary sector performance, Bengaluru has significance difference in terms of contribution to state's GDP. It exhibits polarized economic performance of the city in Karnataka, where it has a share of 29.5% in states GDP and the next three districts (Belgaum, Dharwad and Mysuru) have 2–6% share of the same (Table 2.3). The polarized economic development is also evident in terms of income. Bengaluru has average per capita income of 370,003 which is 128.5% above the state's average; whereas, Bengaluru region, excluding the Bengaluru city has an average per capita income 22% less than state's average.

Bengaluru is the city that received most benefits from economic liberalization started back in 1991. The mounting employment opportunities offered by domestic and multi-national companies, specifically from IT sector has greeted influx of migrants from all over the country. In addition to that, better educational institutes and business prospects have paved the way for people of all skills and economic classes. With that significant share of urban population, Bengaluru is planning to accumulate 20.3 million people in 2031 (The Revised Master Plan of Bengaluru 2017); for that purpose, Bengaluru Development Authority will expand city to another 80 km^2 area. At present, the city is expanding with 500 families and 80,000 m^2 of built-up area per day within its limits and the trend is expected to be same in the next decade (WRI website). Bengaluru Development Authority has 1207 km^2 of area under its jurisdiction, of which 17.63% is under residential land use which is planned to increase to 48.03% by 2031 to accommodate the swelling population of the city. Built-up area is one of the major indicators for assessing level of urbanization along with population densities and other urban facilities. Urban built-up area in Bengaluru Urban District was 37.49% in 2015–16; whereas Mysuru and Hubli Dharwad have 2.61% and 2.67% urban built-up area respectively of their total district area (Table 2.4). Visual observations (Fig. 2.1) also reveal the spatial extent of the urban areas and growth of peripheral townships/settlements in Bengaluru are also noticeably higher in Bengaluru as compared to the other two districts of the state of Karnataka.

2.3   The Push–Pull Paradigm

The densification of population in peripheral area of a city is induced by many push and pull factors. Standard models emphasize on rural-urban migration based on rural push like agricultural modernization and rural poverty, and urban pull factors such as industrialization and urban biased policies to explain urbanization (Jedwab and Vollrath 2016). In the initial state of peri-urbanization, a city gets influx of rural population which can be persuaded by expectation and social networks of migrants and increased investments; whereas, decline in wage differentials can settle down the

Table 2.4 Land use land cover change in one decade in three districts

LULC classes	Area in percentage					
	Bengaluru (U) 2005–06	Bengaluru (U) 2015–16	Mysuru 2005–06	Mysuru 2015–16	Hubli-Dhar 2005–06	Hubli-Dhar 2015–16
Built-up-urban	26.54	37.49	2.03	2.61	1.79	2.67
Built-up-mining	1.14	1.54	0.01	0.16	0.00	0.15
Agriculture-plantation	20.29	14.83	6.20	9.16	0.31	1.19
Wetlands/waterbodies/river/stream/canals/lake/ponds	4.57	4.93	3.56	4.86	1.43	1.41
Built-up-rural	1.01	3.16	1.53	2.24	1.52	1.83
Agriculture-crop land	22.07	26.46	48.09	63.17	75.23	79.64
Agriculture-fallow	10.50	1.25	19.27	0.45	9.21	2.91
Forest-deciduous/EG/semi-EG	6.61	3.75	15.68	13.21	7.12	5.47
Forest-scrub Forest	1.08	1.83	0.4	1.85	1.63	3.33
Forest-forest plantation	0.56	0.43	1.38	0.25	0.29	0.15
Barren/unculturable/wastelands/grazing etc.	5.63	4.33	1.39	2.06	1.47	1.26

Source Calculated based on data from Indian Earth Observation Visualization (https://bhuvan-app1.nrsc.gov.in)

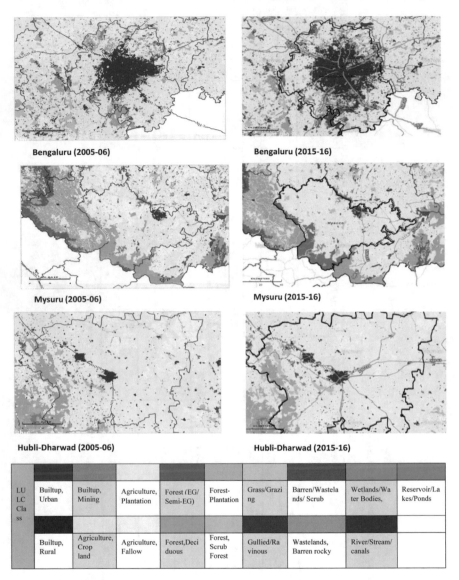

Fig. 2.1 Land use land cover maps of three major urbanized districts of Karnataka. *Source* Indian Earth Observation Visualization (https://bhuvan-app1.nrsc.gov.in)

outmigration from rural areas (Hao 2012). Counter-urbanization caused by cultural aspirations and different socioeconomic drivers which can be grouped under three categories viz.- affordability, accessibility and amenity also make urban people settle down in peri-urban areas (Butt and Fish 2016). Those factors that drive peri-urban transitions are at various levels of scale, which intensify or inhibit peri-urban change (Rauws and de Roo 2011). Presence of labor-intensive industries becomes the most influence factor for many migrants, rather than the driving factors from the place of origin (Wijaya et al. 2018). The scarcity of space in city core incidentally make them to settle in urban hinterlands. Employment generating infrastructure development such as markets (Bryant and Charvet 2003), increased mobility and road networks (Sieverts 2003) in city peripheries contribute as pull factors for people from urban core as well as rural areas to migrate and accelerate peri-urban transformations. The rural-urban fringe transformation in India is due to increasing impact of the urban area on nearby villages, whereas in developed countries, it is due to diffusion of population. In Indian context, the assumption is that the peri-urban areas formation is related to 'push factors', such as a deteriorating environment, creating a strong influence on these areas beyond the traditional city limits (Thirumurthy 2005). Study in Uttarakhand, India (Hoffmann et al. 2019) shows that socioeconomic opportunities such as education, employment, urban facilities act as pull factors for rural population to migrate to cities.

2.4   Rural-Urban Interface and Concept of Peri-Urbanization

Urbanisation brings about changes in land-use patterns by transforming urban–rural linkages, and leads to loss of forest and degradation of environmental quality, because of concentrating population on less land area even though urban land expansion is prevalent (Dutta 2012). Availability and easy accessibility of land in the city surrounding is a major enabling factor for city expansion creating a broader transitional peri-urban area. Along with natural peri-urbanization in convenient surroundings, development of planned satellite towns has become an integral part of urban expansion that eventually merges with the city clusters. Gradually these outgrowths contribute to expansion of city limits. This expansion and ensuing land-use change is a direct driver of damage or destruction of ecological resources of such areas.

According to Vizzari (2011), urban periphery represents a complex landscape because of its proximity and mutual dependence with the cities and rural areas. Organisation for Economic Co-operation and Development (OECD) in its report on peri-urban agriculture (OECD 2003) defines peri-urban areas, which states:

"The term "peri-urban area," cannot be easily defined or delimited through unambiguous criteria. It is a name given to the grey area which is neither entirely urban nor purely rural in the traditional sense; it is at most the partly urbanized rural area. Whatever definition may be given to it, it cannot eliminate some degree of arbitrariness (OECD 2003)." Scholars (Yang and Ye 2020) explained peri-urban interface as socioeconomic system with interactions of rural-urban flows at the periphery of a city.

Australia is the only country with an official definition of peri-urban area. This is defined as "inner regional area" by Australian Bureau of Statistics Remoteness Index.

Although there is formal recognition of peri-urban areas with bureaucratic classification, the fluidity of community experience and landscape change in peri-urban
regions are not addressed (Beilin and Wilkinson 2015). Peri-urban interface has been
termed as the edge of the city, urban fringe or spatial transition zone. Morello (1995)
described that the peri-urban gradient can be understood by examining the mosaic of
"natural" ecosystem, "agricultural" ecosystem and "urban" ecosystems involved in
the flow of materials and energy required in urban and rural areas (as cited in Allen
et al. 1999). Peri-urban area is the most active zone of urbanization that is affected by
urban core (Douglas 2012; Frenkel and Ashkenazi 2008; Huang et al. 2006). Iaquinta
(2001) refers to peri-urban areas as transitional zone or zone of interaction with both
rural and urban activities juxtaposed. These zones have dynamic landscapes with
rapid anthropogenic modifications (Douglas 2012), defined by population, built-up
areas, infrastructure, administrative boundary etc., (Tacoli 1998) and host a wide
variety of functions (Busck et al. 2006). The type fringes are also determined by
the size and function of the bordering city the services (agricultural produce, forest
produce, labour, water, recreational) and resources it is providing towards the city.

Ravetz et al. (2013) explains peri-urbanization in terms of differential industrialization; in developed and old industrial countries, peri-urban zone functions as
social, economic and spatial change, whereas in developing countries it is a zone of
chaotic urbanization leading to sprawl. It has a higher dependence on rural activities
for wealth and employment (in agriculture, mining and fisheries) in developing countries than developed countries (Lynch 2004), thereby exerting a greater pressure on
the biophysical landscape. Simon (2008) analysed the key peri-urban interface issues
in poor and middle-income countries pertaining to both biophysical and socioeconomic landscape change. The major concerns listed include the rate and scale of
land use and land cover change, loss of agricultural land, intensive market-oriented
farming of high value crops, unsustainable use and depletion of both renewable and
non-renewable resources, detrimental health and environmental impacts of wastes,
particularly landfills.

2.5 Fate of India's Rural-Urban Interfaces

Massive peri-urbanization happens when a country approaches towards the advance
stages of development. India has experienced early suburbanization and stagnancy
in metropolitan areas is partially due to the push of firms and workforce out of the
city core, which is facilitated by land management practices (World Bank 2013) and
environmental policy. As a result, there is proliferation of industries, expansion of
urban areas with conversion of agricultural land (Pandey and Seto 2015; Moghadam
and Helbich 2013; Mallupattu and Sreenivasula Reddy 2013; Fazal 2000) and change
in the livelihood patterns of peri-urban communities (Narain 2009). Research shows
(Purushothaman et al 2016) that Indian peri-urban areas indicate a certain gradient
from rural to urban only in case of indicators pertaining to environmental externalities. The controlling factors for peri-urbanization and rural-urban linkages vary

from one to another peri-urban area, which set the mandate to assess the diversified sustainability concerns in peri-urban landscapes. Ramachandra et al. (2012), Ramachandra and Aithal (2013), Reddy and Reddy (2007), Goel (2011), Hackenbroch and Woiwode (2016), Vij and Narain (2016), Dupont (2007), Narain and Nischal (2007), Narain (2009) and Dutta (2012) studied peri-urban areas in different Indian context. Those studies include the spatio-temporal dynamics of urbanizing landscape, top down policy and planning focus, population dynamics, urban edge expansion and envelopment at the cost of permanent crops and pastures, material and service flow livelihood enhancement etc. The available studies reiterate the need of integration of various sectors and advocate for bottom up approach for urban expansion planning where the opinions of various stakeholders would be accounted.

Indian peri-urban areas have weak basic services, and metropolitan peripheries fare poorly on access and quality (World Bank 2013) and failing to generate any of the gains in income, happiness, and mobility that the US, Brazil and China have experienced (Chauvin et al. 2017). The doomed state of environment in peri-urban areas is explained as a result of official neglect and non-recognition to award urban civic status by Saxena and Sharma (2015). Marshal and Randhawa (2014) attribute the poor state of peri-urban areas to institutional obscurity, unplanned growth, lack of infrastructure and environmental degradation. The World Bank report (Urbanization Beyond Municipal Boundary 2016) looks at the role of public policy in potential productivity gains of urbanization with an emphasis on land management policy. It suggests that integrated improvement of land policy, infrastructure and connectivity can help in obtaining optimum benefits from the expanding urban areas in India.

With the understanding so far, we can opine that governance is the prime aspect to consider for addressing the problems associated with peri-urban development. Incompetency of rural institutions in tackling urban-growth-born issues, non-inclusion of peri-urban areas in municipal boundaries, and insensitivity and inadequate knowledge on ecological resources have been emerged as complications in this regard.

2.6 Bengaluru's Experience in Rural-Urban Interface

Shaw (2005) discussed the environmental dimension of spreading urbanization. The findings included problem of increased environmental vulnerability due to solid wastes in the peri-urban areas, which can be managed by improved governance and local initiatives. Another important environmental concern identified is the shifting of polluting manufacturing industries to the periphery of the cities. This shift is encouraged by available low-cost land, accessibility to unorganized workforce, weak implementation of environmental regulations due to lack of awareness in the city periphery (Kundu 2011). Growing environmental concerns generally lead to shifting of large and polluting industries outside the city limit, thereby concentrating industrial activities and settlements of working population as well in the periphery. The migrant workers are also able to find an affordable shelter in the peripheral zone of the city and along with a job in the industries and can avail a less troublesome commute to the workplace even if it is in the city core (Kundu et al. 2002).

The rapid deterioration of ecosystem services in Bengaluru started long back in early 2000. A study by Sudhira et al. (2003) shows from 2002 to 2003, open spaces including parks, lakes and water tanks reduced from 4.4 to 2.5%. Among 262 historically documented lakes in 1961, only 33 lakes could be identified as waterbody from satellite imagery in 2008 (Fernando 2008). Mundoli et al. (2018) studied 31 lakes in peri-urban areas of Bengaluru and findings show that four are converted to other land use and eight are degraded due to pollution, dumping of sewage and garbage and growth of invasive species. Another study (Nagendra et al. 2012) on urbanization and vegetation in Bengaluru reveals increased vegetation clearing and fragmentation; however, the vegetation in the city core is in comparatively stable. Impact of economic growth in terms of environmental degradation is much apparent in city peripheries with changes in ecosystem services and land use (Sudhira and Nagendra 2013).

Water quality of surface water bodies have depleted in the study area along the rural-urban transect and not suitable for human consumption (Dhanush and Devakumar 2019). Research in the same area shows that pesticide use in particular has a negative effect on bee abundance, which spills over to neighbouring plots up to a distance of four km. Results also reveal that bee diversity decreases with continuing intensive plot management (Steinhubel and Cramon-Taubadel 2019). Another study shows reduction land area for cereals, pulses, oilseeds, plantation crops and forest crops; whereas, fruit crops, vegetables, fodder crops and flower crops either have same or increased land under cultivation in rural, transition and urban zones. Results of another study show that satellite towns and road infrastructure are the main channels by which urbanization drives agricultural transition (Steinhübel 2018). Depletion of organic content and micronutrients along the gradient of urbanization, low to high, has also been observed; where conventional crops are found to be helpful in sustaining organic carbon and micronutrients (Kuntoji et al. 2021).

2.7 Looking Forward

Urbanisation brings new opportunities for many, but also results in a drastic increase in the concentration of poverty and environmental degradation in the urban areas and their surroundings, which include the peri-urban-rural landscape. Multi-institutional landscape with overlapped functioning and newly emerging areas of concern with low knowledge and expertise in areas located in rural-urban interface should get policy to overcome the challenges. The Shyama Prasad Mukherji Rurban Mission (SPMRM) launched in 2016 by Government of India aims development of a cluster of villages that preserve and nurture the essence of rural community life with focus on equity and inclusiveness without compromising with the facilities perceived to be essentially urban in nature. Thus, this unique mission too non-inclusive in terms of considering peri-urban areas which are already influenced by urban growth. It has fourteen components under three groups, viz.- livelihoods, services and infrastructure which are planned envisaging a reduced rural-urban divide in terms of economic development. However, it ignores the fact that despite of having relatively better infrastructure, urban areas exhibit inequality, conflict and unemployment; thus, merely providing infrastructures without considering socioeconomic barriers will not build the path to achieve the mission's objective (Singh 2016).

Rapid changes in land use land cover due to residential and commercial expansions that replaces agricultural and natural land uses including common property resources and water bodies, around urban area. Intensification of the transformation is likely to be accelerated by the primate cities like Bengaluru where economic drivers have brought in unpredictable alteration of ecosystems and livelihoods. The interaction between rural and urban is intense in the periphery not only because of the proximity but also because of the demand-supply of the resources. Different processes and levels of urban influences manifest significant social impact that is spatially and temporally varied. These linkages are important not only for their contribution to livelihoods and local economies, but also to acknowledge their role in economic, social and cultural transformation. In countries like India, the alteration of villages into urban centre is very rapid, transforming demographic, economic and biophysical characteristics of the landscape. In this process of transformation, eco-sensitive areas and rural population dependent on natural resources are threatened in many ways, including their livelihoods, socioecological and cultural settings. While discussing the problems of peri-urban areas pertaining to environment, sociopolitical constraints, complex governance, we should not undermine the potential of rural-urban interface in providing resources to meet the urban demand as well as employment to rural population. Thus, the policy initiatives should also consider strengthening and enabling the potential of such areas for by managing jeopardized state of environment.

2.8 Identifying Agroecosystems in Rural-Urban Interface

Discussion in both the previous chapters has already recognized the need of research to understand FMV in Indian transitional agroecosystems. While talking about transitional systems, rural-urban interface has been proved as the most vulnerable landscape to change and environmental degradation. With that understanding to pursue comprehensive learning of field margin vegetation in rural-urban interface of Bengaluru, a transect in the northern periphery of the city has been selected. Phytosociological assessment, socioecological survey, analysis of remote sensing data have been conducted in the selected transect which was divided into three zones namely- urban, transitional and rural. The study area delineated for this research is located in the northern part of Bengaluru which covers 141,563 ha of land (Fig. 2.2). In a study by Hoffmann et al. (2017), in the same transect, they have modified the Urban–Rural Index (Schlesinger 2013) to develop a Survey Stratification Index based on the proportion of built-up area around the villages and their distance from the city

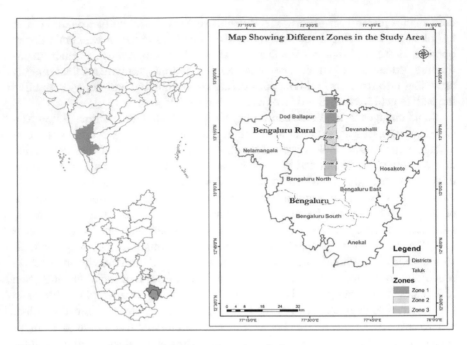

Fig. 2.2 Location of study area in northern Bengaluru, India

core. The zones are selected based on the level of urbanization and random selection of sample villages.

The Bengaluru northern transect covers north taluk (a subdivision of a district) of Bengaluru Urban district and Doddaballapur and Devanahalli taluks of Bengaluru Rural district. The Study area is located between latitude 13°24′ 24.86″ N to 13°3′ 39.414″ N and longitude 77°33′ 42.701″ E to 77°36′ 28.573″ E. About 148 villages are covered in the study area out of which 71 villages are covered in Bengaluru Urban district and 77 villages are covered in Bengaluru Rural district.

Bengaluru is located in the south-east of south Indian state of Karnataka and agro-climatic zone is eastern dry region. Agriculture plays a significant role overall socio-economic growth in Karnataka. The study area is characterised by relatively high temperature during summer. The transect covers area under Bengaluru Urban and Bengaluru Rural districts' jurisdictions. The justification for selection of study area was on the basis of rapid growth in northern part of Bengaluru because of establishment of IT parks and International Airport. The phytosociological study was carried out in six different villages along the northern transect of Bengaluru. The socioecological study was carried out in six different villages along the northern transect of Bengaluru two villages in each, rural, transition and urban regions, respectively.

2.9 Climate, Forest and Physiography

The northern transect of Bengaluru is in South Interior Karnataka meteorological zone. The study area gets most of its precipitation during south west monsoon. The description of meteorological parameters is presented in Table 2.5. The annual average rainfall of the study area is 836 mm. The summer season begins by middle of February and extends up to the end of May or beginning of June followed by Southwest monsoon season extending up to the end of September when the weather is cool and damp. The Northeast or the retreating monsoon season is the period from October to November, while the cold season is from December to the middle of February. The Normal temperature in the Northern transect of Bengaluru ranges from 16 °C to 34 °C. The average maximum temperature is 24.5 °C. January is the coldest month of the year when the temperature is 16 °C. Temperature begins to rise rapidly from the latter half of February. The maximum temperature attains the highest value of about 32–34 °C in the month of April. The area has ranges of hills which are actually spurs of the Western Ghats stretching north-wards with peaks. Bengaluru northern transect situated at the height of 914 m above sea level. The study area is classified under Eastern Dry Zone of agroclimatic classification.

The forest vegetation types in Karnataka include tropical evergreen, semi-evergreen, moist deciduous, dry deciduous, thorny scrub, shola and mangrove. Karnataka's forests are repository of rich biodiversity at the level of gene, species and ecosystem. The state of Karnataka has a complex landscape of species-rich climax forests, secondary forests, pastures, fields and fallows, with corridors of rivers, streams, gorges and ridges, as also a long coastline and marine stretch that are responsible for the rich faunal and floral composition of the state. The study area comes under Bengaluru Forest Circle of Department of Forest, Karnataka. It covers parts of two forest divisions- Bengaluru Urban and Bengaluru Rural. The geographical area of the Bengaluru Rural division is 226,000 ha. This division has about 20,427.49 ha of forest area. It has one sub-division, namely, Doddaballapura sub-division and consists of four ranges namely Doddaballapura, Devanahalli, Hosakote, and Nelamagala. The headquarters of Bengaluru Rural division are situated at Savakanahalli forests, Devanahalli taluk. Most of the forest blocks of Devanahalli and Hoskote ranges have been brought under plantations, mainly with *Eucalyptus*. Forests of Doddaballapura and Nelamangala ranges have some diversity of species such as *Dodonaea viscosa, Lantana camara, Cassia fistula, Stereospermum chelonoides, Albizia amara, Gardenia gummifera, Acacia catechu, Prosopis juliflora*. The forests in the study area are very much scattered all over and are located in small pockets. The forests are very open very open and stunted, are known to possess great variety in terms of plant and animal species. The vegetation in these landscapes, being categorized as scrub, itself demands for a special status for conservation. The hardiness of various species to withstand the vagaries of nature in the form of extreme heat and drought conditions renders the bio-diversity in area a specialty and calls for a concerted effort to inventories, assess the extent and preserve/conserve the components, for the benefit of future generations (Singh 1988).

Table 2.5 Monthly rainfall (mm) and temperature (°C) pattern of Bengaluru Rural and Bengaluru Urban districts (100 years' average)

Bengaluru Urban

Parameters	Jan	Feb	Mar	Apr	May	Jun	Jul	Aug	Sep	Oct	Nov	Dec	Average
Rainfall	3.34	4.32	9.85	51.20	127.30	80.72	94.75	97.16	122.70	164.39	66.33	16.58	838.66
Minimum temperature	16.02	17.46	19.57	21.44	21.25	20.43	19.94	20.00	19.73	19.53	18.06	16.41	19.16
Average temp	22.12	24.03	26.38	27.84	27.11	24.92	23.99	24.13	24.35	24.10	22.88	21.76	24.47
Maximum temperature	28.27	30.66	33.19	34.28	32.99	29.47	28.09	28.28	29.04	28.69	27.77	27.15	29.82

Bengaluru Rural

Parameters	Jan	Feb	Mar	Apr	May	Jun	Jul	Aug	Sep	Oct	Nov	Dec	Average
Rainfall	2.53	3.89	10.48	52.12	125.79	83.31	95.77	98.84	130.03	161.45	58.42	11.54	834.17
Minimum temperature	15.94	17.48	19.56	21.45	21.17	20.39	19.92	19.99	19.72	19.51	17.97	16.22	19.11
Average temp	22.10	24.08	26.43	27.93	27.09	24.80	23.85	23.98	24.28	24.09	22.84	21.65	24.43
Maximum temperature	28.29	30.77	33.31	34.48	33.04	29.29	27.80	28.01	28.91	28.69	27.77	27.12	29.79

Source: India Water Portal and India Meteorological Department (IMD)

2.10   Land Use and Agriculture

The total geographical area of Bengaluru Rural district is 229,519 ha, of which, 100,226 ha is the net sown area and 103,446 ha is gross cultivated area which work out to a cropping intensity of 103%. The district has 11,322 ha under forest. Although 6 rivers originate from Nandi hills (Chickkaballapur district) which were flowing with full water in Bengaluru Rural district in about 5 decades back, have all been reduced to small seasonal streams in view of urbanization, deforestation and encroachment. The soils of the district are mainly red sandy loams which cover an area of 2.19 lakh ha. Doddaballapur and Nelamangala taluks have 100% red sandy loam soils while Devanahalli and Hosakote taluk have 99% of red sandy loam soils and only 1% clayey lateritic soils.

No major river run through Bengaluru Urban district, though the Arakavathi and South Pennar cross paths at the Nandi Hills, 60 km to the North. The river Vrishab-havathi, a minor tributary of Arkavathi arises within the city at Basavanagudi and flows through the city. Bengaluru has a handful of freshwater lakes and water tanks. Of the total geographical area of the district 2.17 lakh ha, the net sown area is only 45,246 ha and the gross cropped area is 46433 ha. It works out to the cropping intensity of 103%. The district has 5055 ha under forest area.

Bengaluru rural district, although agrarian in nature and due to proximity of Bengaluru city, agriculture and allied sectors has taken second priority. Out of the total population of 9.91 lakhs, nearly 73% of the population is residing in Rural areas. Most common crops grown in the district are finger-millet, maize, field bean, red-gram and groundnut. Important horticulture crops grown are mango, grapes, papaya, sapote, vegetable crops such tomato, capsicum, cucumber and floriculture (cut flowers). To supplement the income of the rural folks apart from cultivation of Agriculture and Horticulture crops, Animal Husbandry, Poultry and Sericulture are other important activities. Table 2.6 shows pattern of change in area (in Ha) under different crops in Bengaluru Urban and Bengaluru Rural districts.

Traditionally the agriculture system in Bengaluru Urban district is dependent on rainfall. The district has about 75.5% of its area under agriculture crops, of which cereals occupy a major area 29,791 ha, followed by pulses (4100 ha), oilseeds (1167 ha). The district has 1030 ha (2.2%) under mulberry cultivation and recorded a cocoon production of 1033 tons. Live-stock farming is one of the important occu-pations in this district. The district has a total live-stock population of 3.2 lakhs, of which 1.4 lakhs cows, 8453 buffaloes, 77,302 sheep, 44,725 goats and 18,114 pigs. Figure 2.3 presents the pattern of change in area under different agricultural crops and gradual decrease in crop areas under almost all the categories has been noticed. Among the few crops with increasing area under cultivation, maize in Bengaluru Urban District is notable. In Bengaluru Urban district, horticulture is another impor-tant sector of the district covering an area of 10,345 ha (22.3%) and can be seen with almost stabilized land area under total fruits and vegetables, and a minor increase in area under vegetable crops (Fig. 2.4). With conversion of agricultural land and simul-taneous increase in demand to feed the swelling urban population, crops and cropping

Table 2.6 Pattern of change in area (in Ha) under different crops in Bengaluru Urban and Bengaluru Rural districts

Crops	Bengaluru Rural district			Bengaluru Urban district		
	1998–99	2006–07	2016–17	1998–99	2006–07	2016–17
Food grains	210,495	127,365	52,406	69,051	30,193	20,821
Sugarcane	1018	2232	39	67	27	7
Fruits	14,940	26,854	9488	4005	4254	2774
Vegetables	4414	7724	5523	2898	2191	3756
Total food crop	232,950	167,886	70,417	76,465	37,191	27,800
Total oilseeds	47,397	32,779	3558	6212	3680	1519
Coffee	16	35	2	0	7	3
Fodder crops	5372	6019	3555	2950	1039	746
Green manure	809	21	0	0	0	0
Other non-food crops	36,600	44,994	30,032	9882	14,215	7106
Total non-food crop	90,202	83,856	37,147	19,044	18,941	9374
Total cropped area	323,152	251,742	107,564	95,509	56,132	37,174
Area sown more than once	25,360	8755	3267	13,177	1874	2749
Net area sown	297,792	242,987	104,297	82,332	54,258	34,425

Source Land use statistics information system

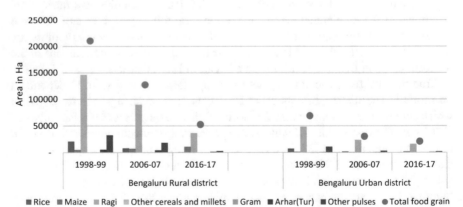

■ Rice ■ Maize ■ Ragi ■ Other cereals and millets ■ Gram ■ Arhar(Tur) ■ Other pulses ● Total food grain

Fig. 2.3 Decadal Change in crop area under cereals and pulses in Bengaluru Rural and Urban districts

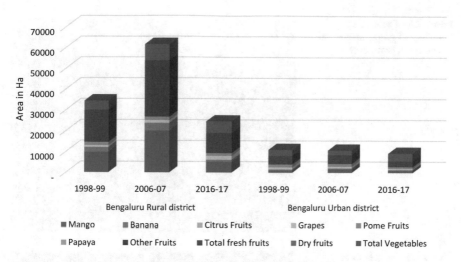

Fig. 2.4   Area under fruit and vegetable crops in Bengaluru Rural and Bengaluru Urban districts

patterns have been changing in both the districts. Some views of different types of farms in the rural-urban interface (including of Bengaluru Rural and Bengaluru Urban districts) of Bengaluru city are presented in photo plates (Photos 2.1, 2.2, 2.3, 2.4, 2.5 and 2.6), where the changing agroecosystems with variances in field margins are apparent (Fig. 2.5).

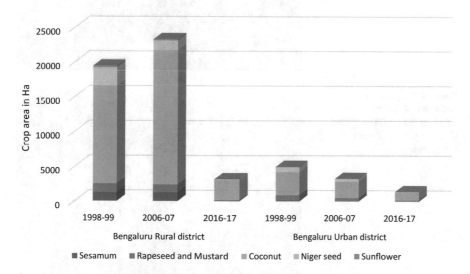

Fig. 2.5   Area under oil crops in Bengaluru Rural and Bengaluru Urban districts

Photo 2.1 Typical farm
boundary of grape farming in
the northern transect of
Bengaluru

Photo 2.2 Farm boundary
of a fallow land in the
northern transect of
Bengaluru

Photo 2.3 Typical farm
boundary of finger millet
(ragi) farm in transition zone
of the northern transect of
Bengaluru

Photo 2.4 Typical farm
boundary of finger millet
(ragi) fields in the rural zone
of the northern transect of
Bengaluru

Photo 2.5 Farm boundary
of a floriculture plot in the
northern transect of
Bengaluru

Photo 2.6 Typical farm
boundary of market-oriented
vegetables farm in the
northern transect of
Bengaluru

2.11　Demography and Socioeconomics

Bengaluru Rural district has 4 taluks, 17 hoblies, 98 gram-panchayats and 1052 villages. In view of the rapid urbanization and availability of employment in Bengaluru city, the rural population has gradually migrated to city. However, the remaining population in rural areas is taking up allied occupations of horticulture, dairying and sericulture on large scale. The total population of the district is 990,896. Of this, 509,172 are male and 481,724 are females. SC population is 213,700 (21.6%) the ST population is 52,903 (5.3%). Further, Doddaballapur taluk has the highest population of 299,594 (30.2%) and highest urban population (101,048).

In Bengaluru Urban district, out of the total population of 96.2 lakhs, 50.2 lakhs are male and 46.0 lakhs are female, while the SC population is 11.98 lakhs and the ST population is 1.9 lakhs (As per 2011 Census). The SC & ST population works out to 14.4% of the total population. The literacy rate is 88 per cent and the sex ratio is 944 females for 1000 males. The percentage of SC population is highest in Bengaluru East taluk (24.4%), followed by Anekal Taluk (21.3%) and Bengaluru North (19.0%). Percentage of ST population is highest in Bengaluru North taluk (3.6%).

Selected Sample Village Profiles

The randomly selected villages for the study are *Konaghatta, Kundana, Heggadihalli, Muddenahalli, Arekere and Singanayakanahalli*. Table 2.7 describes socioeconomic parameters of selected villages. The selected villages fall in Doddabalpur taluk of Bengaluru Rural district and Bengaluru North taluk of Bengaluru Urban district. The total population in Singanayakanahalli as per 2011 census, was 2877 with 18.94% of Schedule caste population and 2.88% of schedule tribes population. The total population of Arakere is 772 with 30.44% of Schedule caste population. The total population of Kundana, Konaghatta, Muddenahalli & Heggadihalli was 1378, 3663, 334 & 1222 with 30.04%, 19.41%, 0.90% and 3.11% of Schedule caste population and with 15.89% 2.07%, 19.16% and 2.29% of schedule tribes population respectively.

Table 2.7 Description of socioeconomic parameters of selected villages

Village Panchayaths	No. of households	SC (%)	ST (%)	Literacy rate (%)	% of agricultural population	% of agricultural labour
Singanayakanahalli	2877	18.94	2.88	67.36	5.28	9.73
Arakere	772	30.44	0.00	50.91	19.04	11.40
Kundana	1378	30.04	15.89	71.63	24.09	6.82
Konaghatta	3663	19.41	2.07	67.59	26.67	14.52
Muddenahalli	334	0.90	19.16	66.47	30.24	0.60
Heggadihalli	1222	3.11	2.29	68.09	14.81	5.73

Source Census of India (2011)

Literacy is important as it determines the workplace and life options of the population and is a determinant for adoption of improved management practices and technologies. In 2011, Kundana village has the highest literacy percentage among the selected villages with 71.63%. Singanayakanahalli, Konaghatta, mussenahalli & Heggadihalli the literacy percentile is 67.36%, 67.59%, 66.47% & 68.09% respectively. Arakere village has the lowest literacy rate among the sample village with 50.91%.

It is observed that the youth are migrating from rural to towns and cities and some out of the village for higher education and employment. This and the possible aging of the farming sector which most countries are experiencing has implications for the future shape of agriculture and management of natural resources. As per 2011 census, Singanayakanahalli village has 5.28% of agricultural population and 9.73% of agricultural labour. The agricultural population in Arakere, Kundana, Konaghatta, Muddenahalli & Heggadihalli are 19.04% 24.09% 26.67%, 30.24% & 14.81% respectively. The agricultural labour in the Arakere, Kundana, Konaghatta, Muddenahalli & Heggadihalli are 11.40%, 6.82%, 14.52%, 0.60% & 5.37% respectively.

References

Allen A, da Silva N, Corubolo E (1999) Environmental problems and opportunities of the peri-urban interface and their impact upon the poor. Document produced for the Research Project Strategic Environmental Planning and Management for the Peri-urban Interface. Development planning unit, University College London. https://www.ucl.ac.uk/dpu/pui

Beilin R, Wilkinson C (2015) Introduction: Governing for urban resilience

Bryant C, Charvet JP (2003) Introduction: the peri-urban zone: the structure and dynamics of a strategic component of metropolitan regions/Introduction: la zone periurbaine: structure et dynamiquesd'unecomposantestrategique des regions metropolitaines. Can J RegNal Sci 26(2–3):231–50

Busck AG, Kristensen SP, Præstholm S, Reenberg A, Primdahl J (2006) Land system changes in the context of urbanisation: Examples from the peri-urban area of Greater Copenhagen. GeografiskTidsskrift-Danish J Geogr 106(2):21–34

Butt A, Fish B (2016) Amenity, landscape and forms of peri-urbanization around Melbourne, Australia. Conflict and change in Australia's peri-urban landscape 7–27

Census of India (2011) Provisional population totals. New Delhi: Office of the Registrar General and Census Commissioner

Chandramouli C (2011) General R. Census of India 2011. Provisional Population Totals, New Delhi. Government of India:409–13

Chauvin JP, Glaeser E, Ma Y, Tobio K (2017) What is different about urbanization in rich and poor countries? Cities in Brazil, China, India and the United States. J Urban Econ 98:17–49

Dhanush C, Devakumar AS (2019) Consequences of urbanization on surface water bodies water quality along the rural-urban and transition zones of Bengaluru. Int J Curr Microbiol App Sci 8(12):2014–30

Desa U (2018) Revision of World Urbanization Prospects. May 2018

Douglas I (2012) Peri-urban ecosystems and societies: transitional zones and contrasting values. In The peri-urban interface: 41–52. Routledge

Dupont V (2007) Conflicting stakes and governance in the peripheries of large Indian metropolises—an introduction. Cities 24(2):89–94

Dutta V (2012) Land use dynamics and peri-urban growth characteristics: reflections on master plan and urban suitability from a sprawling north Indian city. Environ Urban ASIA 3(2):277–301

Eswar M, Roy AK (2018) Urbanisation in Karnataka: Trend and spatial pattern. J RegNal Dev Plan 7(1)

Fazal S (2000) Urban expansion and loss of agricultural land-a GIS based study of Saharanpur City, India. Environ Urban 12(2):133–149

Fernando V (2008) Disappearance and privatisation of lakes in Bangalore. Dialogues, proposals, and stories for global citizenship. https://base.d-p-h.info/en/fiches/dph/fiche-dph-7689.html

Frenkel A, Ashkenazi M (2008) Measuring urban sprawl: how can we deal with it? Environ Plan B: Plan Des 35(1):56–79

Goel N (2011) Dynamic planning and development of peri-urban areas: a case of Faridabad city. India J 8:15–20

Grimm NB (2008) The changing landscape: Ecosystem responses to urbanization and pollution across climatic and societal gradients. Ecol Environ 6(5): 264–272

Hackenbroch K, Woiwode C (2016) Narratives of sustainable Indian urbanism: the logics of global and local knowledge mobilities in Chennai, South Asia. J Interdiscip Multidiscip Res (14)

Hao L (2012) Cumulative causation of rural migration and initial peri-urbanization in China. Chin Sociol Rev 44(3):6–33

Hoffmann EM, Jose M, Nölke N, Möckel T (2017) Construction and use of a simple index of urbanization in the rural-urban interface of Bangalore, India. Sustainability. 9(11):2146

Hoffmann EM, Konerding V, Nautiyal S, Buerkert A (2019) Is the push-pull paradigm useful to explain rural-urban migration? A case study in Uttarakhand, India. PloS one 14(4):e0214511

Huang B, Shi X, Yu D, Öborn I, Blombäck K, Pagella TF, Wang H, Sun W, Sinclair FL (2006) Environmental assessment of small-scale vegetable farming systems in peri-urban areas of the Yangtze River Delta Region, China. Agric, Ecosyst & Environ 112(4):391–402

Iaquinta D, Drescher A (2001) More than the spatial fringe: an application of the peri-urban typology to planning and management of natural resources. In presentation to the conference 'Rural-urban encounters: Managing the environment of the Peri-urban interface

Jaysawal D, Saha S (2014) Urbanization in India: An impact assessment. Int J Appl Sociol 4(2):60–5

Jedwab R, Vollrath D (2016) The urban mortality transition and the rise of poor mega-cities. Manuscript, University of Houston

Kundu A (2006) Globalization and the emerging urban structure: Regional inequality and population mobility, India. Soc Dev Rep, Oxford, New Delhi

Kundu A (2011) Trends and processes of urbanization in India. Urbanization and emerging population issues - 6. London, UK: IIED & UNFPA

Kundu A, Pradhan BK, Subramanian A (2002) Dichotomy or continuum: Analysis of impact of urban centres on their periphery. Econ Polit Wkly 14:5039–5046

Kuntoji A, Subbarayappa CT, Sathish A, Ramamurthy V, Mallesha BC (2021) Effect of different levels of nitrogen and zinc application on growth, yield and quality of maize in rural and peri-urban gradients of southern transect of Bengaluru. J Pharmacogn Phytochem 10(1):1562–8

Lynch K (2004) Rural-Urban interaction in the developing world. Routledge Perspective on Development, Routledge. ISBN 1134513984, 9781134513987

Maiti S, Agrawal PK (2005) Environmental degradation in the context of growing urbanization: a focus on the metropolitan cities of India. J Hum 17(4):277–87

Mallupattu PK, Sreenivasula Reddy JR (2013) Analysis of land use/land cover changes using remote sensing data and GIS at an urban area, Tirupati, India. The Scientific World Journal

Moghadam HS and Helbich M (2013) Spatiotemporal urbanization processes in the megacity of Mumbai, India: a markov chains-cellular automata urban growth model. Appl Geogr 40:140–149

Morello J (1995) Manejo de Agrosistemas Peri-Urbanos, Centro de Investigaciones Ambientales, FAUD, UNMdP, Mar del Plata

Mundoli S, Manjunatha B, Nagendra H (2018) Lakes of Bengaluru: the once living, but now endangered peri-urban commons

Nagendra H, Nagendran S, Paul S, Pareeth S (2012) Graying, greening and fragmentation in the rapidly expanding Indian city of Bengaluru. Landsc Urban Plan 105(4):400–406

Narain V (2009) Growing city, shrinking hinterland: land acquisition, transition and conflict in peri-urban Gurgaon, India. Environ Urban 21:501–512

Narain V, Nischal S (2007) The peri-urban interface in Shahpur Khurd and Karnera, India. Environ Urban, 261–273

OECD (2003) OECD Environmental indicators: Development, measurement and Use. PARIS: Environmental performance and information division, organization of economic co-operation and development.

Pandey B, Seto KC (2015) Urbanization and agricultural land loss in India: Comparing satellite estimates with census data. J Environ Manag 148:53–66

Purushothaman S, Patil S, Lodha S (2016) Social and environmental transformation in the Indian peri-urban interface—emerging questions. Azim Premji University

Ramachandra TV, Aithal BH (2013) Urbanisation and sprawl in the Tier II City: metrics, dynamics and modelling using spatio-temporal data. Int J Remote Sens 3:66–75

Ramachandra TV, Setturu B, Aithal BH (2012) Peri-urban to urban landscape patterns elucidation through spatial metrics. Int J Eng Res Dev 2(12):58–81

Randhawa P, Marshall F (2014) Policy transformations and translations: lessons for sustainable water management in peri-urban Delhi, India. Environ Plan C Polit Space 32(1):93–107

Rauws WS, de Roo G (2011) Exploring transitions in the peri-urban area. Plan Theory & Pract 12(2):269–84

Ravetz J, Fertner C, Nielsen TS (2013) The dynamics of peri-urbanization. In Peri-urban futures: Scenarios and models for land use change in Europe. Springer, Berlin, Heidelberg, pp 13–44

Reddy VR, Reddy BS (2007) Land alienation and local communities: case studies in Hyderabad-Secunderabad. Econ. Polit Weekly, 3233–3240

Roy A (2002) City requiem, Calcutta: Gender and the politics of poverty. Minneapolis, MN: University of Minnesota Press

Saxena M, Sharma S (2015) Periurban area: a review of problems and resolutions. Int J Eng Res Technol 4(09):2278–0181

Schlesinger J (2013) Agriculture along the urban-rural continuum. A GIS-based analysis of spatio-temporal dynamics in two medium-sized African cities

Shaw R, Das A (2018) Identifying peri-urban growth in small and medium towns using GIS and remote sensing technique: A case study of English bazar urban agglomeration, West Bengal, India. Egypt J Remote Sens Space Sci 21(2):159–72

Sieverts T (2003) Cities without cities: An interpretation of the Zwischenstadt. Routledge

Simon D (2008). Urban environments: Issues on the peri-urban fringe. Annual Rev Environ Resourc 33:11–19

Singh NP (1988) Flora of Eastern Karnataka, vol 1. Mittal Publications

Singh SK (2016) Releasing into conflict zones: exploring the spatial drivers of urban insecurity and its impact on the reentry of offenders in New Orleans. Doctoral dissertation, Tulane University, Payson Center for International Development

Steinhübel L (2018) Somewhere in between towns, markets, and neighbors: Agricultural transition in the rural-urban interface of Bangalore, India. Discussion Papers

Steinhubel L, Cramon-Taubadel SV (2019) Somewhere in between towns, markets, and jobs—Opportunity costs of agricultural intensification in the rural-urban interface

Sudhira HS, Nagendra H (2013) Local assessment of Bangalore: Graying and greening in Bangalore–impacts of urbanization on ecosystems, ecosystem services and biodiversity. InUrbanization, biodiversity and ecosystem services: Challenges and opportunities. Springer, Dordrecht, pp 75–91

Sudhira HS, Ramachandra TV, Jagadish KS (2003) Urban sprawl pattern analysis using GIS

Tacoli C (1998) Rural-urban interactions: a guide to the literature. Environ Urban 10(1):147–66

Taubenböck H, Wegmann M, Roth A, Mehl H, Dech S (2009) Urbanization in India–Spatiotemporal analysis using remote sensing data. Comput Environ Urban Syst 33(3):179–88

Thirumurthy AM (2005) Periurban deliverable 2: Socio-economic conceptual frame work (WP2). Div Urban Syst Dev, Chennai

Vij S, Narain V (2016) Land, water and power: the demise of common property resources in periurban Gurgaon, India. Land Use Policy 50:59–66

Vizzari M (2011) Spatial modelling of potential landscape quality. Appl Geogr 31(1):108–18

Wijaya HB, Kurniawati H, Hutama ST (2018) Industrialization impact on worker mobility and land use in peri urban area (case study of Semarang district, Indonesia). InIOP Conference Series: Earth and Environmental Science 123(1):012037. IOP Publishing

World Bank (2013) Urbanization beyond municipal boundaries: nurturing metropolitan economies and connecting peri-urban areas in India. Directions in development. World Bank, Washington D.C

Yang Y, Ye L (2020) Peri-urban development. Oxford Bibliographies in Urban Studies

Chapter 3
Agroecosystems in Rural-Urban Interface

3.1 Trade-offs in Peri-Urban Agroecosystems

It is understandable that anthropogenic land use land cover change is an inevitable worldwide phenomenon with increasing intensity. It has extensive implications on forest and water resources, biogeochemical cycles, climate and food production systems. Those changes affect the available habitats for microbes to big mammals. As a response to environmental stress and changing natural resources, local population directly dependent on ecosystem goods for their livelihoods are forced to shift to other livelihood options affecting their social and economic constructions. In general, the concept of urbanization is associated with growth. However, growth is considered as activator for economic wellbeing but with unavoidable negative externalities. With the expanding urban core, the fringe area is also not static; it gradually gets converted into rural-urban periphery and then merges with urban centre. Rural urban fringe is a mix of both urban and rural features. The characteristics of such area are neither identical with urban or rural, but correspond to both, and some special characteristics specific to such transition zones.

The vibrant processes involving land and demography make the peri-urban areas transitional and hotspot of services to support the city with multiple functions starting from agricultural production, residential and recreational (Busck et al. 2006). Thus, the peripheral areas undergo manifold alterations (physiological, socioeconomic, cultural and functional) with consequences like uncertain and complicated pattern of land use land cover change that may be undesirable for ecological as well as social sustainability. The fringe villages supply the resources required for the city expansion and in return, their natural and semi-natural (agricultural land) are deteriorated or completely transformed to other land use. The interaction between rural and urban is intense in the periphery not only because of the proximity but also because of the demand-supply of the resources. Different processes and levels of urban influences manifest significant social impact that is spatially and temporally varied. These linkages are important not only for their contribution to livelihoods and local economies, but also to acknowledge their role in economic, social and cultural transformation.

In consort with government induced programmes, urbanization plays a substantial role in changing the characteristics of agroecosystems. Land use land cover change, market oriented agricultural production, change in food habits etc., upshot in the process. Rapid urbanisation and the expansion of the megacities act as major driving force for the loss of farmland biodiversity (Tscharntke 2005). Land use changes are related to the rate of urbanisation along with the transition from an agriculture-based to a non-agricultural economy (Sen 2015). The rapid change in land use has resulted in the decline of biodiversity (Wood 2000). The drivers for agricultural land use change include demographic, economic, technological, institutional, and socio-cultural and location factors (Jasper 2015).

Land use land cover change for human use leads to transformation of native ecosystem into agricultural land or built-up land like housing (Collinge 1996) and such changes are evident as leading cause many detrimental environmental impacts (Bella and Irwin 2002), for e.g.- of biomass loss (Henson 2005). The lack of proper planning and development in the rural-urban periphery contributes to the jeopardy of urbanization in Indian states (Sandhu 1996). The process of formation and characteristics of peri-urban areas in developing countries including India are different from that of developed countries. Towns and cities have also been growing haphazardly with consequent rise in environmental degradation.

Bunting (2007), Johanna et al. (2009), Agrawal et al. (2003), Brook et al. (2001), FAO (2005) and others have focused on studies related to peri-urban agriculture in India with special reference to waste water use in agriculture and aquaculture, pollutant residues in food products, impact of air pollution, food security and livelihood enhancement in peri-urban areas of various cities in India. Study in fringe areas of Delhi reveals that although the villagers have been exposed to prolonged urban influences, land continues to be an integral part of their lives specially in terms of acquiring their livelihoods (Mallik and Sen 2011). The peri-urban areas of sprawling cities experience significant land transformation, due to expansion of the urban core contained within their boundaries (Dutta 2012). The studies also reveal different processes and levels of urban influences that manifest significant social impact with prominent temporal variation.

Sell and conversion of land by farmers in peri-urban areas is massive in Indian context. Abrupt increase in land price caused by construction of highways and industrial infrastructure lead them to either sell or leave agricultural land uncultivated or keeping land just as an asset and take up alternative livelihoods. Establishment of borewell for urban water supply in agricultural land of peri-urban area of Bengaluru is evident (Ravishankar et al. 2018). In peri-urban areas, industrial pollution has also contributed towards decrease in agriculture. Forceful purchase of land at low price by real-estate agencies from socially backward communities is another crucial dimension of peri-urban growth. Lack of ownership which may be because of continuous growth of infrastructures in both rural and urban areas is one major problem

(Aijaz 2019). The actors in peri-urban interface exhibit a strong role and the land transformation is guided by the "alliances" and "conflicts" among the actors (Fazal 2012). Thus, a multiplicity of changes with varied causative factors exist and play role of different degrees in the transformation of peri-urban areas. Therefore, the transition zones or peri-urban areas need enough attention to be studied in detail in the context the drivers of development which affects the extent and nature of development. Considering the huge economic returns, farmers have been selling their agricultural lands to non-agricultural land use, consequently increasing the employment pressure on urban regions by the agrarian populations (Rathee 2014). In order to achieve food sufficiency for the growing population, the land under agriculture has to be increased; whereas it is being reversed due to the process of urbanisation.

3.2 Socioecological Household Survey and Assessment of FMV

The transect was divided into three parts on the basis of population density, distance from the city core and extent of subsistence agriculture and land transformation. List of villages selected in northern transect of Bengaluru for the Field Margin Vegetation (FMV) study are presented in Table 3.1. Household questionnaire survey was conducted for socioecological assessment. Informed individuals were interviewed and then sample for stratified survey has been selected. Farming households were considered as strata for sample selection. More than 30% of households have been surveyed from each village. The questions were grouped into household profile, general socioeconomic information, land owning and use details, agricultural practices and production, livestock, types of vegetation on field margin, use and TEK of FMV, reasons for change in FMV. Information have been gathered for two points of time—past (in 1990s) and present to enable analysis for assessing pattern of landscape change and subsequent change in FMV.

To get basic information about the current status of species diversity and structure of field margin vegetation, plot level data were collected from each village. Ten plots in each village were randomly selected from the sample of household survey so

Table 3.1 Details of study area for the field margin vegetation in the northern transect of Bengaluru

S. no.	Type	Name of the village	No. of households	Village area (ha)
1	Rural	Heggedahalli	242	269
2		Muddenahalli	81	72
3	Transition	Kundana	217	412
4		Konaghatta	659	727
5	Urban	Singanayakanahalli	357	467
6		Arekere	148	313

that coordinated field data, historical information and perception of farmers can be collected. Precaution was taken to cover all types of major croplands for the detailed socioecological study of field margin vegetation.

3.3 The Changing Field Margins in Rural-Urban Interface of Bengaluru

Urbanisation result in the process of spatial diffusion from centre to peripheral areas, transforming most of the rural areas into urban characteristics. Significant and rapid growth in urbanization in and around Bengaluru has brought about substantial changes in agriculture, natural vegetation including forest and field margins, and water resources. Especially, the productive agriculture lands of the transition areas of Bengaluru have been undergoing transformations which affect the quality, diversity, and scale of agricultural crop production and spatial patterns.

The phytosociological assessment reveals that in our study area, along with economically important exotic tree species such as *Eucalyptus*, Silver Oak etc., farmers have retained and managed fruit bearing species such as mango, guava, jackfruit, mud apple and custard apple in their field boundaries to some extent. The density of those species has been found higher as compared to other species naturally occurring species in field boundaries of all the six villages. Research by Patil et al. (2018) in the same study area shows transitions in agroecological and sociocultural setting in tune with urbanisation gradients are evident in peri-urban Bengaluru. In this context, it is essential to understand how field boundaries can act as a repository of species diversity and change along the urban gradient which has tremendous importance in terms of sustainable agroecosystem and conservation. The same study shows that the transitions are from subsistence and locally or regionally integrated farming systems (for example, ragi–pulses) to increasingly commercialised and globally linked crop value chains (vegetables and flowers). These changes are expected to have adverse impact on diversity, multifunctionality and resilience of agroecosystems.

The FMV of the studied plots show maximum of 52 species in rural section of the transect followed by urban (40 species) and the least with 31 species in transition zone. The transition zone, although has half of the farm plots under cereals and pulses, it has shown low richness of species diversity in the FMV (Table 3.2). It can be attributed to maximum utilization of farm area, cultivation of grapes (11.9%) and lawn grass (7.4%) which is associated with complete clearing of vegetation from field boundaries.

Annexures show the extensive list of species present in different groups of crops and percentage of fields with tree in the field margins. Cereals and pulses have shown tree-FMV in 41.67, 61.90 and 51.14% of fields in rural, transition and urban sections of the transect. FMV in Vegetable crop fields of transition zone has been low with only 27% of farms having tree in FMV and grape fields have no FMV in the transition

Table 3.2 Percentage of plots under different crops

Zones	Cereals and pulses	Vegetables and flowers	Plantation crops	Grapes	Lawn grass
Rural	48.98	34.69	6.12	10.2	0.0
Transition	50	26.19	4.76	11.9	7.14
Urban	35	17.5	17.5	25	5

Source Authors' estimation

zone. Grapes cultivation has been found as one of the most potential crops to remove field boundary vegetation. The household survey also reported that in rural zone of the transect, 33.8% of the households stated that expansion of their agricultural land has engulfed the field boundary areas which compelled them to remove FMV (Fig. 3.1). On the other hand, 8–9% of households in urban and transition zones removed FMV while expanding their crop area (Photos 3.1 and 3.2)

The socioecological survey conducted for the selected farms in the transect reveals decrease in number or elimination of a few tree species in their farm boundaries in last 30 years in all the three zones. Such FMVs undergoing diminishing abundance include nine species. Tamarind has been reported to be removed from field boundaries of farmers of all the three sections of the transect along the urban gradient. It is noticeable that all the diminishing species are native species with immense socio-cultural and economic importance.

The percentage of households in the studied sample has only 5% in the landscape who stated as increase in agricultural land in the current decade (Fig. 3.2). 49% of the households' agricultural land area has reduced in recent years. The noticeable changes in the transect in terms of agricultural activities are found to be governed

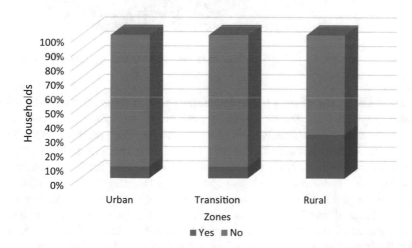

Fig. 3.1 Respondents on FMV area engulfed for agricultural expansion. *Source* Authors' estimation

Photos 3.1 and 3.2 Types of field boundaries in two different zones of the transect (Ragi field with FMV and grape field without any FMV)

Fig. 3.2 Status of change in agricultural land among the farming households. *Source* Authors' estimation

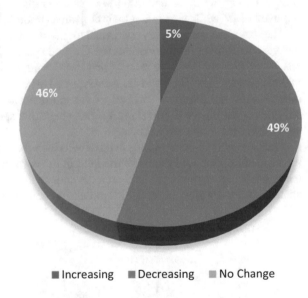

■ Increasing   ■ Decreasing   ■ No Change

by landholdings, land quality, water availability, market price of land and market-oriented crop selection. The changes observed in the transition zone of the study transects have been substantially realised by the famers which has been attributed to socioeconomic changes and reduction of landholdings because of division of ancestral land among brothers. These two reasons together account for 50% of the households (Fig. 3.3). These two causes of agroecological landscape change are followed by lack of water and land sale as major concerns for the farmers. Socioeconomic change and lack of water are stated as major reasons for change in agricultural landscape by the small and marginal farmers (with agricultural land < 5 acres); whereas, the relatively larger farmers have stated market influence and division of agricultural land as major drivers for change.

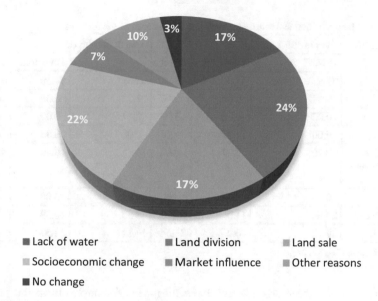

■ Lack of water   ■ Land division   ■ Land sale
■ Socioeconomic change   ■ Market influence   ■ Other reasons
■ No change

Fig. 3.3 Percentage of households stating different reasons for change in agricultural landscape. *Source* Authors' estimation

The number of crops cultivated in 1990s was 43 in the transition zone which has decreased to 17 in the present decade. Market-oriented farming and government induced agroforestry started in the zone prior to 1990s. In recent years, reduction in landholding size because of division, land sale and degradation have led the farmers to alter their crop selection with interest of subsistence farming.

In the present agricultural system, the farmers in the urban zone are taking up ragi and maize as their main crop. Among the surveyed farmers, 57.14% of them cultivates ragi and maize as their main crop and 21.43% of farmers cultivate fruits as their major crop; whereas, 14.29% and 7.14% of the farmers are into horticulture, specifically vegetable and plantations respectively (Table 3.3). Among the farmers in the transition zone, 57.14% cultivates ragi and maize and 14.28% cultivates fruits as their major crop. 24.42% and 7.14% of farmers have plantation and vegetables as their major crop. In rural section, 66.67% of farmers cultivates ragi and maize as their major crop, fruits and coconut plantation is the next major crop with 13.33% and 20% respectively taken up by the farmers.

It is evident that ragi being the traditional staple of the population, has the least decrease in cultivation area which is 11.33% in urban, followed by rural (7.4%) and transition (4.4%). In urban part of the transect, the change in agricultural land use significant with 65 and 22% increase in area under fruit and coconut plantation, and *Eucalyptus* plantation from 1990s to present decade. In transition zone, the farmers have 28% of their agricultural land converted into *Eucalyptus* plantation in the current decade. Among the studied population, pulses barely had any land occupied in 1990s; but now pulses have been introduced among the urban farmers. Paddy, mulberry and

Table 3.3 Change in area under different crops in three zones of the transect

Major crops	Area in 1990s (in %)			Area at present (in %)		
	Urban	Transition	Rural	Urban	Transition	Rural
Ragi/Jowar	51.52	57.06	45.32	45.68	54.55	41.96
Paddy	3.64	0.00	0.00	1.23	0.00	0.00
Other cereals	0.00	0.00	0.00	2.47	0.00	0.00
Mulberry	0.00	12.27	9.94	0.00	0.00	0.00
Plantation (coconut and fruits)	9.70	12.27	0.00	16.05	6.49	40.18
Pulses	0.00	0.00	0.00	4.94	0.00	0.00
Vegetables	23.03	11.04	23.20	13.58	10.39	8.93
Eucalyptus	12.12	0.00	21.54	14.81	28.57	8.93

Source Authors' estimation

sugarcane are the three crops which have disappeared completely in the present decade among the studied population (Table 3.4).

With decrease in average agricultural landholding from 5.4 acres (2.2 ha) per household in 1990s to 2.8 acres (1.13 ha) per household in the present decade, the variety of crops cultivated in the past has significantly decreased. The transition zone of the landscape has already encountered the pressures exerted by urbanization, particularly on land which has been noticed in the form of diminishing area under agriculture. Among all the three zones, transition zone has shown highest growth of built-up area (77.56% change) and horribly alarming decrease in waterbodies

Table 3.4 Number of FMV trees present in agricultural plots of different crops

Crops	Average Number of FMV trees per plot*		
	Rural	Transition	Urban
Cereals/pulses	6.42	7.26	8.92
Flower	4.00	3.67	3.00
Grapes	2	0.00	1.6
Plantations	4.8	4.3	3.0
Vegetables	1.33	1.40	10.00
Fallow	Xx	21.00	Xx
Lawn grass	Xx	0.00	0.00
Total	8.31	6.07	7.67
Grand total	7.39		

Xx—no farm present
Source Authors' estimation (*The numbers given here based on average from the multiple fields. This indicates number of trees not species. The diversity and density varies and linked with the farm practice, crop selection and socio-economic conditions of the farmers)

Table 3.5 Percentage change in area under different land use land cover classes from 1991 to 2018

Change: 1991–2018	Built up	Agriculture	Open forest and plantation	Water body	Scrub land	Barren land
Rural	54.53	−7.15	71.44	−257.43	−38.48	−236.06
Transition	77.56	−25.25	60.05	−771.47	0.66	−194.07
Urban	56.50	−35.48	−27.13	−61.23	18.37	−214.26

Source Authors' estimation from LULCC analysis (details in Chap. 5)

(−771.47% change) from 1991 to 2018 (Table 3.5). This dynamism of the physical landscape parameters is strong enough to state that the transition zone has undergone changes that are detrimental for the agricultural system of the zone. The decrease in area under water body also substantiates the statement by 23% of the surveyed households which mentioned lack of water as the foremost basis for the changing agricultural landscape.

3.4 Socioeconomic Landscape and Direct Economic Benefits from FMV

The modern pattern of urbanization is contributing towards new complexities of rural-urban relationships, predominantly influencing the social and economic structure of peri-urban population. This structural change in socioeconomics of the population bring in change in the agricultural system depending on their changing needs and desire. People in the adjacent rural area either migrate or travel to the peripheral city region to acquire the benefits of new occupations which are apparently better paid, with urban facilities and urban way of living. The cumulative impact of urban expansion is noticeable in various socioeconomic characteristics. In general, the urban influence is found to vary according to the gradient of the city in terms of income, education, occupation, social adhesion, cultural attributes etc.

Figure 3.4 shows the percentage of households under five income categories of agricultural households. In rural zone, 100% of the surveyed households and in urban zone 94% of the surveyed households have agriculture as their primary livelihood option. The urban agricultural households, despite of having higher average land-holdings than other zones, 96% of its population has an annual income less than Rs. 25,000 per annum, whereas the rural farmers are having higher income (46.48% in 1–5 lakh per annum category and 8.45% in more than 5 lakh per annum category) (Fig. 3.5).

Almost half of the surveyed households have experienced decrease in their agricultural landholdings, whereas only 5.08% has an increase in their agricultural land area since 1990s. Among the surveyed households, transition section farmers have small landholdings as compared to rural and urban sections. Urban farmers own larger agricultural lands as compared to other two sections. On an average each

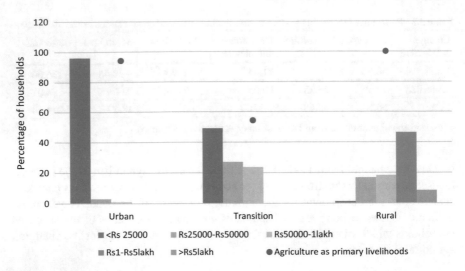

Fig. 3.4 Distribution of households under different income classes in the study transect

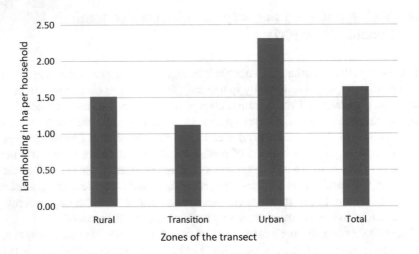

Fig. 3.5 Average landholding per household in the study transect

urban farmer owns 2.31 ha of land; whereas, rural and transition zone farmers have an average of 1.51 ha and 1.12 ha of land per household respectively. Compared to 1990s, the present landholding has decreased by 1.84% in the urban zone. Similarly, 7.19% and 16.33% of decrease in landholdings have been noticed in transition and rural zones respectively.

Field investigations and interactions with farmers revealed that the cash crops they are cultivating now do not require shade hence the many native tree species which farmers used to raise in the field margins are no longer required. Another factor which was found responsible for decline in area of FMV is expansion of main

Photo 3.3 Interface between two farms in the study area

land use for increasing the profit from cultivation of cash crops. The cash crop driven land use change has negatively affected the structure and functional characteristics of field margins and found all across the regions of northern periphery of Bengaluru. Optimum utilization of a field boundary for economic benefit in the transition zone of the rural-urban interface can be seen in Photo 3.3.

Agricultural farm boundaries are places where wild species grow as well as economically and ecologically important species are deliberately planted and conserved. Demographic and economic conditions of the changing landscape determine what the farmers want to plant, retain and use in the field margins. Farm boundaries receives seeds at greater densities than open farms and pastures which makes it species-rich and it needs to be maintained for its numerous socioeconomic benefits (Pandey 2002). Fruit bearing and timber producing trees have been found to play a crucial role in the economy of farm households, especially during crop failure and family distress situation along with supporting livestock, nutritional supply and supporting other faunal diversity.

The tree, shrub and herb species are categorised in 10 use groups that give any type of benefits. Those include fruits, fodder, fuelwood, fibre, ornamental, beverages, timber, leaf litter/compost etc. Among all the uses, use of plants and plant parts as food (fruits and vegetables) includes highest number of species, i.e. 17, followed by fodder, firewood and timber. The food items collected from the field margins are primarily for household consumption with selling of spared produce, mainly coconut, drumsticks

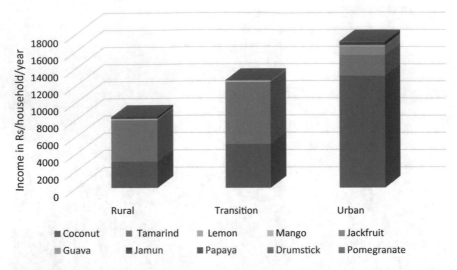

Fig. 3.6 Economic benefit from FMV in terms of foods in Rs per household (HH) per year. *Source* Field survey

and tamarind. Among the food producing trees, coconut, tamarind, papaya, wood apple, jackfruit, mango, custard apple etc., are to mention as prevalent throughout all the three sections of the transect. The average benefits in monetary term per household in the rural-urban interface from the edible FMV (perennial species) products per year is Rs. 12,604; where, the households in the urban zone has higher economic benefit (Rs. 17,070 per household per year) as compared to other two zone (Fig. 3.6).

The survey results have apparently shown decrease in households using field margins vegetation in all the use categories except fodder (Fig. 3.8); however, with respect to quantity used, all the categrories have shown decrease in use of biomass (Fig. 3.10). Although there has been increase in number of households using fodder from 38% in 1990s to 51% at present, the quantities of resources used in all categories have decreased during that period. The increase in fodder harvesting families from field boundary is also another consequence of land scarcity led agricultural intensification which has pushed more families to adopt livestock rearing as secondary livelihood.

The socioecological survey has shown that the farmers stated *Pongamia* spp., *Carissa carandas, Jasminum multiflorum* and farm grasses as the most abundant vegetation in the field boundary where farm grass is new addition as abundant species as compared to *Carissa carandas, Jasminum multiflorum, Leucas aspera, Pongamia* spp. and *Lantana camara* in 1990s. *Eucalyptus*, coconut, mango and silver oak trees are mentioned as less abundant species at the field margin which are economically important. Pomegranate, farm grass, *Eucalyptus*, teak, jackfruit, mango etc., are found to be planted in recent years. The current direct usage of FMV primarily include fodder, crop protection, fruits for both household and commercial purpose. Thus, the semi-natural characteristics of field margin have shifted to plantation crops, thereby adversely affecting the multi-functionality of agroecosystem with diverse non-crop species.

Figures 3.7 and 3.8 show the use of resources from FMV across the transect. The average annual economic benefit from food produces, specifically fruits and nuts accounts for Rs. 12,604 per household. The urban zone, although has less interaction with the field margin vegetation, is having the maximum benefit primarily from

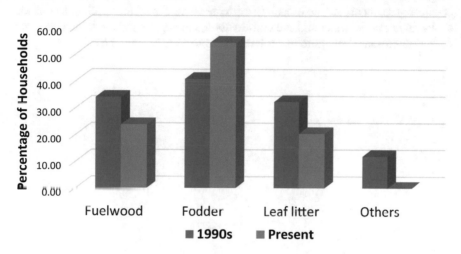

Fig. 3.7 Use of FMV-based resources in peri-urban landscape of Bengaluru, India. *Source* Field survey

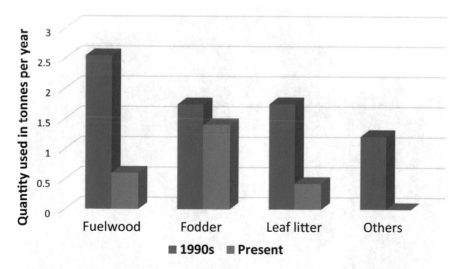

Fig. 3.8 Use of FMV-based resources (estimated raw weight) in peri-urban landscape of Bengaluru, India. *Source* Field survey

coconut. This economic return includes both the benefits- domestic consumption and cash benefit from selling. In transition and rural zones, tamarind provides the highest economic benefit from any food produce of FMV. Both fodder (7240 kg/HH/yr) and fuelwood (1995 kg/HH/yr) collections are higher in transition zone as compared to the others; whereas leaf litter utilization is more in rural zone. The services and ecosystem functions provided by the field boundary and vegetation of the boundary are wide-ranging and complex. Some of the services provided by FMV, which the farmers considered as important but difficult to estimate the values include pollination, shade, shelter to livestock, water and soil conservation, fencing, boundary demarcation etc. A few are presented in photo plates below (Photos 3.4, 3.5 and 3.6).

Photo 3.4 FMV as shade tree

Photo 3.5 FMV as shelter for people and livestock

Photo 3.6 Natural FMV as farm fencing

References

Agrawal M, Singh B, Rajput M, Marshall F, Bell JNB (2003) Effect of air pollution on peri-urban agriculture: a case study. Environ Pollut 126(3):323–329

Aijaz R (2019) India's peri-urban regions: the need for policy and the challenges of governance. ORF Issue Brief, p 285

Bella KP, Irwin EG (2002) Spatially explicit micro-level modelling of land use change at the rural-urban interface. Agric Econ 27(3):217–232

Brook R, Purushothaman S, Hunshal C (2001) Changing frontiers: the peri-urban interface Hubli-Dharwad, India. Books for Change, India

Bunting SW (2007) Confronting the realities of wastewater aquaculture in peri-urban Kolkata with bio-economic modelling. Water Res 41(2):499–505

Busck AG, Kristensen SP, Præstholm S, Reenberg A Primdahl J (2006) Land system changes in the context of urbanisation: examples from the peri-urban area of Greater Copenhagen. Geogr Tidsskr 106(2):21–34

Collinge SK (1996) Ecological consequences of habitat fragmentation: implications for landscape architecture and planning. Landsc Urban Plan 36(1):59–77

Dutta V (2012) Land use dynamics and peri-urban growth characteristics: reflections on master plan and urban suitability from a sprawling north Indian city. Environ Urban ASIA 3(2):277–301

FAO (2005) Fertilizer use by crop in India. Food and Agriculture Organization of the United Nations, Rome

Fazal S (2012) Land use dynamics in a developing economy: regional perspectives from India. Springer Science and Business Media

Henson IE (2005) An assessment of changes in biomass carbon stocks in tree crops and forests in Malaysia. J Trop For Sci, 279–296

Jacobi J, Drescher AW, Weckenbrock P (2009) Agricultural biodiversity strengthening livelihoods in periurban Hyderabad, India. J Agric Urban Entomol 2(22):45–47

Jasper van Vliet HL (2015) Manifestations and underlying drivers of agricultural land use change in Europe. Landsc Urban Plan 133:24–36

Johanna J, Axel WD, Philipp W (2009) Agricultural biodiversity strengthening livelihoods in periurban Hyderabad, India. Urban Agric Magr 2(22):45–47

Mallik C, Sen S (2011) Land dispossession and changes in rural livelihoods: the case of peri-urban Delhi. In: Dikshit JK (ed) The urban fringe of Indian cities. Rawat Publications, New Delhi, India

Pandey DN (2002) Traditional knowledge systems for biodiversity conservation. Organization of the United Nations (FAO) Forestry Paper. FAO, Rome, Italy, pp 22–41

Patil S, Dhanya B, Vanjari RS, Purushothaman S (2018) Urbanisation and new agroecologies. Econ Polit Wkly 53(41):71

Rathee G (2014) Trends of land-use change in India. In: Kala Seetharam Sridhar GW (ed) Urbanization in Asia: governance, infrastructure and the environment. Springer, India, pp 215–238

Ravishankar C, Nautiyal S, Manasi S (2018) Social acceptance for reclaimed water use: a case study in Bengaluru. Recycling 3(4):1–12. https://doi.org/10.3390/recycling3010004

Sandhu RS (1996) Urbanization of India. J Soc Policy, 319

Sen S (2015) Land and livelihood: agricultural land use changes in India. J Geography You, 22–27

Tscharntke TKD (2005) Landscape perspectives on agricultural intensification and biodiversity—ecosystem service management. Ecol Lett 8:857–874

Wood PM (2000) Biodiversity and democracy: rethinking Society and Nature. UBC Press, Vancouver

Chapter 4
Structure and Functions of FMV in Rural–Urban Interface

4.1 Phytosociological Study of Field Margins

The biodiversity in an agroecosystem has a wide range of roles to play. Both species interactions and anthropogenic acts determine the structure and composition of biodiversity in the field boundaries. Being a human-manipulated ecosystem, an agricultural ecosystem mostly supports species diversity direct economic benefits to human population. It largely has an impact on the diversity of naturally occurring both floral and faunal diversity.

A holistic phytosociological assessment of field margin vegetation from selected sample plots was conducted to identify the types of species, their richness and composition use of plant and plant products and traditional knowledge pertaining to each species in the field margin vegetation. Listing of plant species were done according to their habits i.e.—tree, shrub and herb. Further identification and taxonomic classification were done for all species. Phytosociological study deals with the floristic pattern and composition and interactions of species within the community. Field survey for phytosociological assessment was carried in the selected 10 plots per village using standard scientific methods given by various researchers as cited in Nautiyal et al. (2015). The quadrats were laid down along field margin for tree, shrubs and herbs as per the standard vegetation analysis methods (Kershaw 1973; Saxena and Singh 1982; Mehta et al. 1997); shrubs were estimated using a 5 m × 5 m quadrat, herbaceous vegetation was assessed using a 1 m × 1 m quadrat. FMV strips being narrow in breadth, quadrat area of 10m × 10m was calculated from the polygons delineated through remote sensing data. Thus, random areas of same size were selected for studying tree species. The step-by-step approach adopted in this research is given in the flowchart (Fig. 4.1).

For calculating the species composition, abundance and diversity indices at the transect level, the following common variables were used: basal area, relative dominance, relative frequency and relative dominance following Phillips (1959), while the sum of the relative dominance, relative frequency and relative dominance gave the

© The Author(s), under exclusive license to Springer Nature Switzerland AG 2021
S. Nautiyal et al., *Field Margin Vegetation and Socio-Ecological Environment*,
Environmental Science and Engineering,
https://doi.org/10.1007/978-3-030-69201-8_4

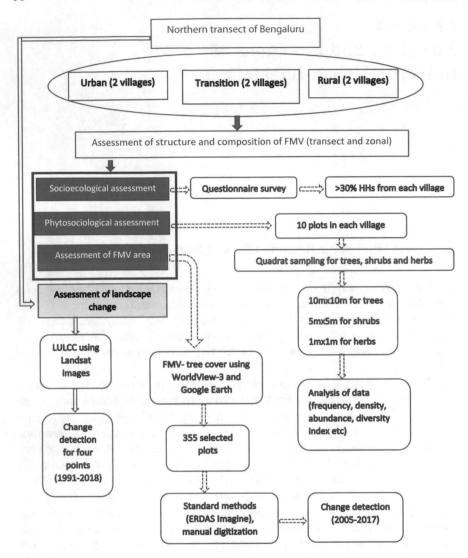

Fig. 4.1 Methodological framework for studying rural–urban interface of Bengaluru describing phytosociological assessment, socioecological survey and land use land cover analysis

importance value indices (IVI) for various species (Curtis 1959). Species richness, concentration of dominance (C^d), evenness and similarity index were also analyzed. For developing the land use land cover change for the study area, GPS data was recorded.

$$\text{Frequency} = \frac{\text{Number of sampling units(quadrates)in which a species occurs}}{\text{Total Number of sampled units studied}} \times 100$$

$$Density = \frac{\text{Total number of individual in all sampling units}}{\text{Total Number of sampled units studied}}$$

$$Abundance = \frac{\text{Total number of individuals in all sampling units}}{\text{Total Number of sampling units of occurence}}$$

$$\text{Basal area} = \pi * (DBH)^2/4$$

$$\text{Relative density}(RD) = \frac{\text{Number of individual of a species}}{\text{Total Number of individual of all species}} \times 100$$

$$\text{Relative Frequency}(RF) = \frac{\text{Number of occurrences of a species}}{\text{Total Number of occurrences of all species}} \times 100$$

$$\text{Relative dominance}(RDo) = \frac{\text{Total basal area of individual species}}{\text{Total basal area of all species}} \times 100$$

$$\text{Importance value index} = RD + RF + RDo$$

Diversity Indices: Diversity is a combination of two factors; the number of species present, species richness and the distribution of individuals among the species are referred to as evenness or equitability. The two most widely used species diversity indices are Shannon and Simpson indices. They are adopted by ecologists to describe the average degree of uncertainty in predicting the species of an individual picked at random from a given community. As the number of species increases, the uncertainty of occurrence also increases along with distribution of individuals, more evenly among the species already present. The Shannon–Wiener species diversity index, when properly manipulated, always results in a diversity value (H′) ranging between 0, indicating a low community complexity and 4, indicating high community complexity.

Species diversity (H′) was computed following the Shannon and Weiner (1963) information index as follows (Eq. 4.1)

$$H = \sum \frac{n_i}{N} \log_n \frac{n_i}{N} \tag{4.1}$$

Where n_i = number of individuals of each species and N = total number of individuals in the sampling plot.

4.2 FMV Species Richness and Diversity Indices

Through the phytosociological study in agriculture field margin in northern transect of Bengaluru a total of 82 plant species including the trees, shrubs and herbs were recorded from different plots within six villages. A total of 30 trees were recorded within trees, whereas 23 and 29 known species recorded in shrubs and herbs communities respectively (Fig. 4.2). The transition zone of the transect has been found as the richest in terms of species diversity among the three zones. The maximum 60 species were recorded in Kundana village followed by 58 species in Konagatta village and both are located in the transition zone of the study region. Village-wise occurrence of FMV species of different families under tree, shrub and herb habits are presented in the Table 4.1. The surveyed plots exhibited rich diversity of families for tree habits (19 families) as compared to shrubs (11 families) and herbs (13 families). Among the herbaceous plants, Asteraceae has highest number of species, i.e. seven species, whereas Malvaceae and Fabaceae are the rich families in shrub and tree categories respectively. Plant species (*Cassia* spp., *Tamarindus* spp., etc.) of Fabaceae family are mostly planted FMV for their economic benefits. Transition zone has a higher densities of plantation crops (Kundana village with 422 *Eucalyptus* trees/ha and Konaghatta with 217 coconut trees/ha), yet this zone has shown a better diversity of FMV species as compared to rural and urban zones of the transect. The shift of natural FMV to semi-natural FMV is quite significant in transition zone for acquiring economic outcomes, where human interference has actually increased the diversity. In urban zone of the transect, interaction between human and FMV was observed low and a subsistence agricultural system is disappearing. It has also resulted in a built-up boundary of the crop field or a plant community going through natural succession. More than half of the tree species documented are native tree. Fruits plants (*Manilkara zapota, Carica papaya* and *Annona squamosa*) and other economically important plants (*Eucalyptus* spp., *Delonix regia, Grevillea robusta*) are abundant throughout the transect. The two most abundant species are *Eucalyptus* spp. and *Cocos nucifera*.

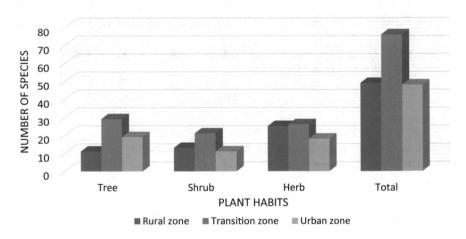

Fig. 4.2 Number of species under different plant types across the study transect

Table 4.1 Family-wise distribution of number of species documented in each village*

	Families	No. of species in each village					
		Heggadihalli	Muddenahalli	Kundana	Konaghatta	Singanayakanahalli	Arekere
1	Arecaceae	1	1	2	1	1	1
2	Fabaceae	3	2	4	4	2	1
3	Myrtaceae	2	2	3	3	2	3
4	Lamiaceae	1	1	1	1	1	1
5	Moraceae	1	1	3	4	1	2
6	Meliaceae	1	1	1	1	1	1
7	Proteaceae	1	1	1	1		1
8	Anacardiaceae	1	1	2	2	1	1
9	Muntingiaceae			1			
10	Mimosaceae			1			
11	Caricaceae			1			
12	Sapotaceae			1			
13	Lythraceae			1			1
14	Annonaceae			1			
15	Musaceae			1			
16	Euphorbiaceae				1	1	
17	Sapotaceae				1	1	
18	Rutaceae				2	1	1
19	Phyllanthaceae						1
	Total-tree species	**11**	**10**	**24**	**21**	**12**	**14**

(continued)

Table 4.1 (continued)

	Families	No. of species in each village					
		Heggadihalli	Muddenahalli	Kundana	Konaghatta	Singanayakanahalli	Arekere
1	Apocynaceae	1	1	3	1	1	1
2	Verbenaceae	1	1	1	1	1	1
3	Asteraceae	3	2	3	3	1	1
4	Malvaceae	4	3	3	5	3	4
5	Solanaceae	1	2	1	3	1	2
6	Amaranthaceae	1				1	1
7	Euphorbiaceae		1	2	1		
8	Nyctaginaceae			1			
9	Rutaceae			1	1		
10	Cactaceae			1	1		
11	Fabaceae				1		
	Total-shrub species	**11**	**10**	**16**	**17**	**8**	**10**
1	Asteraceae	5	7	6	6	3	3
2	Fabaceae	2	2	2	2	2	2
3	Acanthaceae	1	1	1	1		
4	Amaranthaceae	3	3	3	3	3	3
5	Oxalidaceae	1	1	1	1		
6	Rubiaceae	3	4	1	4	2	2
7	Lamiaceae	1	1	1	1		1

Table 4.1 (continued)

	Families	No. of species in each village					
		Heggadihalli	Muddenahalli	Kundana	Konaghatta	Singanayakanahalli	Arekere
8	Euphorbiaceae	2	3	2	3	2	2
9	Cleomaceae	1					
10	Commelinaceae		1			1	
11	Phyllanthaceae		1	1	1		1
12	Apocynaceae			1		1	
13	Araceae			1			
	Total-herb species	**19**	**24**	**20**	**22**	**14**	**14**

Source Field Survey (*The numbers are obtained from the sampling fields (plots). However, the numbers would vary from fields to fields. The species richness is based on the study villages located in each zone. Therefore, the numbers and diversity not necessarily represent the entire zone in which the study villages are located)

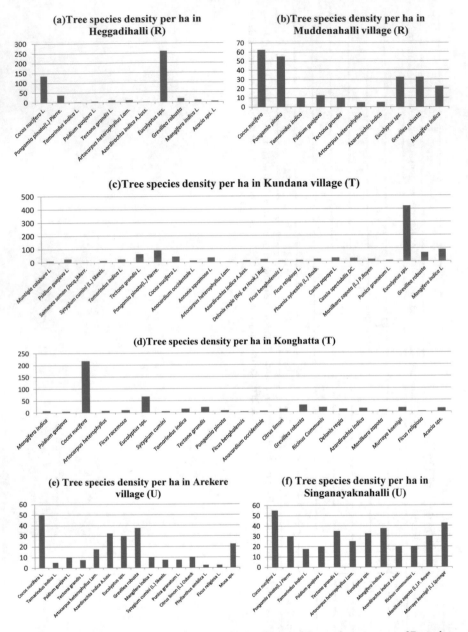

Fig. 4.3 a–f Density of tree species per hectare in six villages of the northern transect of Bengaluru

Tree density of the documented species in all the six sample villages are presented in Fig. 4.3a–f. The results indicate that the structure of field margin vegetation is changing from vegetation with a combination of wild and traditionally grown plant species to economically important exotic plant species.

Relative density of available shrub species in the three zones are presented in Table 4.2. Among the shrub species, weeds are having more relative density, *Parthenium* sp. at the top, followed by *Lantana camara*. Three species of *Sida* plant are available in the study area. Urban part of the transect is completely dominated by weed species belonging to shrub habit, which apparently indicates less maintenance and use of field boundaries by the farmers. The herb species documented in the villages show that Heggadihalli village has the highest number of herbaceous plants (24 species) and urban zone villages-Arekere and Singanaykanahalli have least, i.e. 14 species each. The relative density and relative frequency of FMV-herb species from quadrat survey are presented in Annexures-8 and 9. Among the herbs, *Wedelia chinensis* has the highest value of relative density in both, the villages of rural and transition zones. It covers 9.53–23.55% of density among all herbs in those four villages. The urban zone villages have *Alternanthera sessilis* and *Mimosa pudica* as the most dominant herb species in field boundaries in terms of relative density.

Table 4.2 Relative density (%) of shrub species across the northern transect of Bengaluru

Sl. No	Scientific name	Transition zone	Urban zone	Rural zone
1	*Allamanda cathartica*	1.83		
2	*Abutilon indicum*	3.16		
3	*Calotrophis gigantea*	2.99	3.69	2.03
4	*Thevetia peruviana*	0.99		
5	*Jatropha curcas*	1.99		
6	*Lantana camara*	30.61	16.85	16.45
7	*Parthenium hysterophorus*	38.26	28.17	43.62
8	*Solanum torvum*	4.15	9	7.94
9	*Sida cordifolia*	3.99	9.69	10.72
10	*Sida rhombifolia*	2.16		3.69
11	*Ricinus communis*	1.16		
12	*Bougainvillea spectabilis*	0.99		
13	*Chromolaena odorata*	1.66		4.43
14	*Xanthium indicum*	3.32		1.66
15	*Opuntia dilleni*	1.66		
16	*Sida acuta*		14.08	2.4
17	*Urena lobata*		6.46	
18	*Gomphrena globosa*		12	1.47
19	*Triumfetta rhomboidea*			5.54

Relative density (RD) and Relative frequency (RF) of the species occurring in quadrats laid down in two or more villages are presented in the Table 4.3. Some of the species with low relative frequency of occurrence and are also present in only one of the studied villages include *Delonix regia, Citrus limon, Phyllanthus emblica, Syzygium cumini, Anacardium occidentale, Annona squamosa, Ficus* spp. and *Carica papaya. Cocos nucifera* has the highest relative frequency in five villages except Kundana, where *Eucalyptus* spp. has the highest relative frequency. In terms of relative density, Heggadihalli and Kundana have *Eucalyptus* spp. with highest value among all FMV tree species; whereas, *Cocos nucifera* has highest relative density in other four villages.

Table 4.4 shows Shannon Weiner Diversity indices for all the study villages along with Evenness Index. Singanayakanahalli village scores highest in diversity index (2.43), followed by Arekere village and Kundana village which belong to urban and transition zone respectively observe species diversity index of 2.29. Heggadihalli village field boundaries show a relatively low tree species diversity (H = 1.39) as compared to other villages. The evenness value of Singanayakanahalli is near complete evenness (E = 0.98) which can be interpreted as uniform and equally diverse FMV tree species throughout the village ecosystem.

The Working Plan of Forest Department for Bengaluru Division (2002/03–2011/12) states that the area has an average of 26 trees per hectare of private land. The major species mentioned are *Eucalyptus* (17.2%), Neem (6.3%), Tamarind (1.2%), *Acacia* (2.6%) and *Ficus* (5.3%). In case of FMV, it is found that coconut, *Eucalyptus* and *Pongamia* are the species with highest relative density in the study transect. When the data obtained from quadrat study of field boundary is compared with that, it shows significantly high number of trees in field boundaries, which is 340 per acre or 137 per ha of land in Kundana village of transition zone. It shows the ecological importance of field boundaries in retaining trees in private lands. The percentage tree cover area to the crop area is around 7.5% (results are presented in next chapter). The trees per unit area (number of individuals and population density) across the zones in northern transect of Bengaluru vary from field to field. The density (individual or population) closely linked to the various factors, for example—land holding size, socio-economic condition of farm families, knowledge on importance of conservation and management of FMV etc. When the results of field margin area assessment are correlated, it can be stated that the trees on private lands are concentrated in field margins only. Species-wise basal area, basal area per ha and species-wise volume per ha are presented in the Table 4.5. The results reveal that *Eucalyptus* spp. are the most dominant species in the landscape which is having 1314.11 cubic feet tree volume per acre. The average tree volume per acre of private land in Bengaluru Forest Division is 387.45 cubic feet; whereas a significantly high biovolume of tree stems in field boundaries is apparent from the results of the study, which accounts for 1679.85 cubic feet per acre.

4.2 FMV Species Richness and Diversity Indices

Table 4.3 Relative density and relative frequency of FMV tree species present in six villages of northern transect of Bengaluru

Scientific names	Heggadihalli		Muddenahalli		Kundana		Konaghatta		Arekere		Singanayakanahalli	
	RD	RF	RD	RF	RD	RF	RD	RF	RD	RF	RD	RF
Cocos nucifera	27.27	34.62	25.25	27.66	4.03	5.11	43.07	34.62	21.74	17.14	15.07	10.81
Pongamia pinata	7.58	13.46	22.22	14.89	8.60	10.95	1.49	3.85	–	–	8.22	9.46
Tamarindus indica	0.51	1.92	4.04	6.38	2.15	3.65	2.97	6.41	2.17	2.86	4.79	8.11
Psidium guajava	1.01	1.92	5.05	8.51	2.42	4.38	0.99	1.28	4.35	5.71	5.48	5.41
Tectona grandis	2.02	3.85	4.04	6.38	5.91	8.76	4.46	3.85	3.26	5.71	9.59	8.11
Artocarpus heterophyllus	2.02	3.85	2.02	4.26	0.27	0.73	1.49	3.85	7.61	14.29	6.85	8.11
Azadirachta indica	0.51	1.92	2.02	4.26	1.08	2.92	–	–	14.13	17.14	5.48	6.76
Eucalyptus spp.	52.53	25.00	13.13	8.51	40.86	18.98	13.37	6.41	13.04	5.71	8.90	6.76
Grevillea robusta	3.54	3.85	13.13	8.51	6.45	4.38	5.94	3.85	16.30	5.71		
Mangifera indica	2.02	7.69	9.09	10.64	8.60	9.49	1.49	3.85	4.35	5.71	10.27	9.46
Acacia spp.	1.01	1.92	–	–	–	–	2.97	2.56	–	–	–	–

Table 4.3 (continued)

Scientific names	Hegadihalli		Muddenahalli		Kundana		Konaghatta		Arekere		Singanayakanahalli	
	RD	RF	RD	RF	RD	RF	RD	RF	RD	RF	RD	RF
Ricinus communis	–	–	–	–	–	–	2.97	3.85	–	–	5.48	5.41
Manilkara zapota	–	–	–	–	1.61	1.46	1.49	2.56	–	–	8.22	8.11
Murraya koenigii	–	–	–	–	–	–	3.47	2.56	–	–	11.64	13.51

Table 4.4 Diversity and evenness of FMV tree species in six villages of northern transect of Bengaluru

Parameters	Heggadihalli	Muddenahalli	Kundana	Konaghatta	Arekere	Singanayakanahalli
Shannon Weiner (H)	1.39	2	2.29	2.2	2.29	2.43
Number of species (S)	11	10	24	21	14	12
Maximum diversity possible (Hmax)	2.40	2.30	3.18	3.04	2.64	2.48
Evenness E (II/IImax)	0.58	0.87	0.72	0.72	0.87	0.98

4.3 Manipulation and Management of FMV

Manipulation of vegetation on farm boundaries is normal phenomenon and usually occurs for enhancing economic benefits. During the study period it was observed that not many farmers have been found taking appropriate efforts to manage or maintain the field margin vegetation.

In both urban and transition sections 20% of the fields have boundaries covered with natural plant species. 46.67% of the field margins is covered with semi natural and 13.33% is covered with planted vegetation in urban zone, and 15.56% of fields does not have FMV (Table 4.6). In transition zone, 40% of the boundaries has semi natural, 4.44% is combination of natural and semi-natural and 13.33% is covered with planted vegetation. Transition zone has the highest number of agricultural fields without FMV, i.e. 20%. The rural part of the transect has maximum fields (80%) with semi-natural vegetation boundary. During the household survey, questions have been asked to understand the farmer's perception on impacts of urbanization on the change of FMV in their farms. The responses are anonymous stating that there is no impact of urbanization on FMV. Although the physical and socioeconomic landscapes have been changing as a result of urbanization, no direct and tangible impact have been realised by the farming population (Table 4.7).

Among the surveyed households, it is observed that *Bauhinia* is the species found abundant in the urban part of the transect i.e., 7.14% of plots among the study plots. *Sesbania grandiflora* (7.14%) and Banyan (20%) are the tree species found abundant in the transition and rural sections respectively. *Pongamia* is the species found abundant in all the three sections. Coconut is the planted tree species found abundant in all the three sections. Study in Doddabalapur taluk (Nalina et al. 2017) shows *Parthenium, Mimosa pudica, Pongamia* spp., *Lantana camera, Eupatorium odoratum* and some grasses are the major natural vegetation apart from dominant preferred plant species. Field survey in the study villages reveals that although,

Table 4.5 Basal area and stem volume of FMV species in transition zone of northern transect of Bengaluru

Scientific name	Avg. height in ft	No. of individuals /acre	Avg. GBH in ft	Basal area /tree (sqft)	Basal area (sqft)/acre	Tree volume (cubic ft/acre)
Muntingia calabura	14.67	4	2.34	0.46	1.82	11.22
Psidium guajava	11.02	9	4.41	1.55	13.95	64.58
Samanea saman	48	1	8.74	6.09	6.09	122.71
Syzygium cumini	27.91	4	5.27	2.21	8.86	103.84
Tamarindus indica	28.39	8	5.47	2.38	19.06	227.31
Tectona grandis	47.4625	22	2.58	0.53	11.69	233.08
Pongamia pinnata	24.0125	32	2.52	0.50	16.15	162.92
Cocos nucifera	31.4075	15	3.57	1.01	15.20	200.55
Anacardium occidentale	20.645	4	2.74	0.60	2.39	20.72
Annona squamosa	13.32	12	2.40	0.46	5.48	30.66
Artocarpus heterophyllus	25	1	2.95	0.69	0.69	7.29
Azadirachta indica	23.73	4	3.04	0.74	2.95	29.39
Delonix regia	25	7	6.04	2.91	20.37	213.84
Ficus benghalensis	113.5	2	10.09	8.10	16.21	772.58
Ficus religiosa	32	4	8.67	5.98	23.94	321.71
Phoenix sylvestris	21.2	7	2.28	0.41	2.89	25.71
Carica papaya	15.29	10	1.78	0.25	2.52	16.20
Cassia spectabilis	14.54	9	3.18	0.81	7.26	44.32
Manilkara zapota	20.86	6	2.38	0.45	2.70	23.68
Eucalyptus spp.	46	152	1.85	0.34	51.29	1314.11
Grevillea robusta	30	32	1.73	0.27	8.75	79.90
Mangifera indica	23.55	24	3.08	0.76	18.17	125.18

Table 4.6 Types of vegetation on field boundaries in different zones in the northern transect of Bengaluru

Zones	Natural	Semi natural	Planted	Natural and semi natural	No FMV
Urban	20.00	46.67	13.33	4.44	15.56
Transition	20.00	40.00	13.33	6.67	20.00
Rural	6.67	80.00	6.67	2.22	6.67

Table 4.7 Percentage of
different types of FMV
present in the three sections
of the northern transect (n =
60), Bengaluru

Types of FMV	Urban	Transition	Rural
Natural	7.14	7.14	20.00
Semi natural	50.00	78.57	60.00
Planted	57.14	78.57	33.33

Source Field survey

Lantana camara is an invasive species growing in field boundaries, it barely invades the crop area. Farmers stated that they prefer to retain it on the boundary as it works as a protective barrier. In addition, *Lantana* also acts as natural pest control. Spider webs in those shrubs are very common which help in reducing the insects that are harmful to the crops. The density of *Lantana* varies from 182 individuals/ha (in urban zone) to 511 individuals/ha (in transition zone). It is the second most abundant shrub in field margins after *Parthenium* sp.

4.4   Traditional Ecological Knowledge on Field Margin Vegetation

Traditional ecological knowledge refers to a cumulative body of knowledge, practice and belief, evolving by adaptive processes and passes through generations by cultural means (Berkes 1999); knowledge basebuilt not by experts but by resource users (Berkes 2004) and is specific to a given culture or society (Warren and Rajasekaran 1993) and it focuses on conceptualization and interaction of local culture with biotic and abiotic environment (Gadgil and Berkes 1991). The knowledge is guided by the importance and usage of resources by the indigenous people or local communities and perceptions they acquire through generations. Development of traditional knowledge is a dynamic process that changes with the availability of resources and the demands of local communities (Becker and Ghimire 2003). The description of knowledge varies with actors of the same socioecological system along with geographical differences. Traditional ecological knowledge has an important role in understanding ecological systems and has the ability to support scientific programs in different context (Naidoo and Hill 2006). TEK involves overall information on histories associated with locally available wildlife, cultural norms for land management and resource allocation and management (Becker and Leon 2000; Becker and Ghimire 2003).

The importance of traditional ecological knowledge (TEK) is well established within natural resource management and sustainability science (Zent and Maffi 2009). Science can help mobilize traditional knowledge through preparing methods for obtaining, assessing and presenting as well as preparing inventory of traditional ecological knowledge (Warren 1996). A combination of scientific knowledge and TEK can play a positive role in sustaining biodiversity, nature's services and subsequently sustain the cultural norms of a community (Becker and Ghimire 2003).

Traditional societies have a holistic view of the ecosystem and the social system where relationship with nature is based on coexistence (Ramakrishnan 2005). Thus, the traditional systems help the communities to develop adaptive strategies (Berkes and Jolly 2002) to cope with any change in the biotic or abiotic environment and sustainable use of resources.

The diverse functions of field margins vary with types of agroecosystems and need to be protected with locally strategized conservation efforts to deal with socioeconomic and biophysical landscape change. Local self-management of natural habitat and conservation involving public support demand use of traditional knowledge (Pilgrim et al. 2007). Research on role of TEK with respect to socioeconomic and environmental change be vital in contributing for formulation and implementation of strategies which are policy relevant (Boafo et al. 2016). Use of traditional knowledge in managing ecosystem services are wide-ranging. Berkes et al. (2000) studied various international literature on traditional knowledge pertaining to management of ecosystem processes and functions. Study revealed that traditional knowledge is pertinent to diversity of such concerns including multiple species management, resource rotation, succession management, landscape patchiness management, and other ways of responding to and managing pulses and ecological surprises. Thus, conservation of field margin vegetation for enhancing services of agroecosystems can be strengthened through systematic study and documentation of TEK.

Agricultural farm boundaries are places where wild species grow as well as economically and ecologically important species are deliberately planted and conserved. Demographic and economic conditions of the changing landscape determine what the farmers want to plant, retain and use in the field margins. Fruit bearing and timber producing trees have been found to play a crucial role in the economy of farm households, specially during crop failure and family distress situation along with supporting livestock, nutritional supply and supporting other faunal diversity. Farm boundaries receives seeds at greater densities than open farms and pastures which makes it species-rich and it needs to be maintained for its numerous socioeconomic benefits (Pandey 2002).

The uses of different plant species under all the three habits-tree, shrub and herb are classified into ten use categories (Fig. 4.4). Among all the uses, use of plants and plant parts as food (fruits and vegetables) includes highest number of species, i.e. 17, followed by fodder, firewood and timber. The food items collected from the field margins are primarily for household consumption with occasional selling. Among the food producing trees, coconut, tamarind, papaya, wood apple, jackfruit, mango, custard apple etc., are to mention as prevalent throughout all the three sections of the transect. Traditional knowledge on use of the plants and plant parts includes a variety of practices and processes; preparation of pickle (*Mangifera indica*), beverages (*Phoenix sylvestris*), organic manure (*Crotalaria pallida*), insect repellent (*Azadirachta indica*) and a wide range of traditional medicines and culinary ingredients. Other traditional knowledge includes knowledge on best season for harvesting and use, identifying the usable plant and plant parts, making furniture and storages etc.

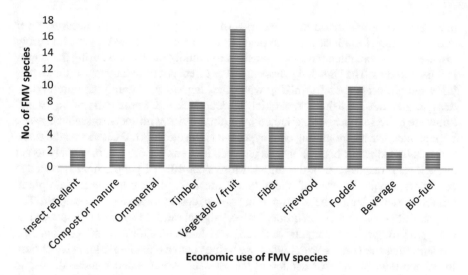

Fig. 4.4 Number of plant species from used to acquire various economic benefits in the northern transect of Bengaluru

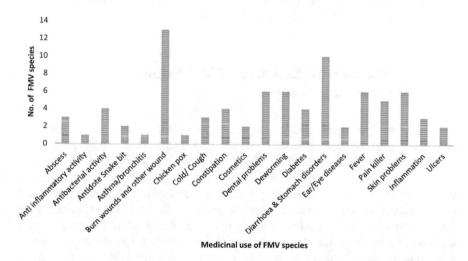

Fig. 4.5 Use of FMV species for medicinal purposes in the northern transect of Bengaluru

The medicinal uses of field margin vegetations are assessed as wide ranging; starting from cold-cough to chicken pox, bronchitis etc. In addition to the medicinal plants that the farmers acquire from home gardens, FMV too provide ingredients for basic healthcare in traditional way. The uses of FMV species for medicinal purposes are listed under 20 broad groups of diseases and ailments (Fig. 4.5). Highest number of species of FMVs, i.e. 13 species are used for different types of wounds including cut and burn, followed by 10 species for curing diarrhoea and stomach disorders, six species each the categories of dental problems, skin diseases, fever and deworming. Details of communities' knowledge on using FMV plants for curing diseases are documented through extensive community interaction during the empirical field studies. The knowledge base involves direct use of plant parts, extraction of juice and oil, proportion for mixing with other ingredients, storing and preserving, dosage, methods for using, preparation of decoction etc. Some parts of traditional knowledge are found to be explained by communities with modern medical basis; for e.g., increase haemoglobin, cure hypoglycaemia etc. It is indicative that the traditional knowledge has been enhanced with modern knowledge and it is likely to get strengthened or changed in the future. More than 40% of people hold knowledge on different types of uses of plants like drumstick, black plum, wild date plum, tamarind, guava, lemon, betel palm, bitter gourd, castor, queensland's hemp etc. The respondents possess more traditional knowledge about tree species (21 species) as compared to species belonging to shrubs and herbs categories. This provides the evidence that the species with higher use values also have a larger knowledge base in the community. In urban section of the transect (Arekere and Kundana), despite of the changing agroecology because of transitional socioecological and physical landscape, farmers are managing the plant species in their field boundaries, more specifically the fruit bearing species such as *Mangifera indica, Manilkara zapota and Musa* spp. which show higher abundance (>3) in the urban part of the transect.

4.5 Way Forward for Enhancing FMV Value and Preserving TEK

This study provides a preliminary understanding on the importance of vegetation on field boundaries with respect to their use and knowledge possessed by the farmers. It is seen that traditional structure of boundary vegetation has been replaced by economically important exotic species, which is more prevalent in rural and transitional sections of the transect where policy induced agroforestry practices are more extensive. Farmers are well aware of the importance of the traditional species in their farmland; but the care should be taken to restrict replacement of those species for achieving economic goals. Documentation of communities' knowledge is essential to prevent biopiracy which may deprive the community in future from their rights and benefits through their knowledge base. Commercialization of knowledge and resource, if done through Intellectual Property Rights, original knowledge holders should be able to get the benefits by appropriate benefit sharing mechanisms. Traditional knowledge systems are area and community specific, and largely not replicable. Swamy (2017) emphasized that documentation of knowledge should not be capitalistic to create monopolization by introducing IPR which has the threat to compress the social and cultural context of the knowledge. Policies should be formulated to encourage agricultural production as a part of the ecosystem which will enable to conserve the non-crop biodiversity at the field margin and their associated knowledge. In this approach, identification and valuing of ecosystem services of all the components of agroecosystem should be carried out.

References

Becker CD, Ghimire K (2003) Synergy between traditional ecological knowledge and conservation science supports forest preservation in Ecuador. Glob Ecol Conserv 8(1):1

Becker CD, Leon R (2000) Indigenous institutions and forest condition: lessons from the Yuracare. McKean GM, Ostrom E (eds) People and forests: communities, institutions, and the governance. MIT Press, Cambridge, Massachusetts, USA 163–191

Berkes F (1999) Role and significance of 'tradition' in indigenous knowledge: Focus on traditional ecological knowledge. Indigenous Knowledge and Development Monitor, Netherlands

Berkes F (2004) Rethinking community-based conservation. Conserv Biol 18(3):621–30

Berkes F, Jolly D (2002) Adapting to climate change: social-ecological resilience in a Canadian western Arctic community. Ecol Conserv 5(2)

Berkes F, Colding J, Folke C (2000) Rediscovery of traditional ecological knowledge as adaptive management. Ecol Appl 10(5):1251–62

Boafo YA, Saito O, Kato S, Kamiyama C, Takeuchi K, Nakahara M (2016) The role of traditional ecological knowledge in ecosystem services management: the case of four rural communities in Northern Ghana. Int J Biodiversity Science, Ecosystem Serv Manag 12:24–38. https://doi.org/10.1080/21513732.2015.1124454

Curtis JT (1959) The vegetation of Wisconsin: an ordination of plant communities. University of Wisconsin Press

Gadgil M, Berkes F (1991) Traditional resources management systems. Resour Manag Optim 8(3–4):127–141

Kershaw KA (1973) Quantitative and dynamic plant-ecology, 3rd edn. ELBS and Edward Arnold Ltd., London

Mehta JP, Tiwari SC, Bhandari BS (1997) Phytosociology of woody vegetation under different management regimes in Garhwal Himalaya. J Trop For Sci 10:24–34

Naidoo R, Hill K (2006) Emergence of indigenous vegetation classifications through integration of traditional ecological knowledge and remote sensing analyses. Environ Manage 38(3):377–387

Nalina CN, Anilkumar KS, Shilpashree KG, Babu N, Sudhir K, Natarajan A (2017) Inventory and mapping of land resources for land use planning through detail soil survey coupled with remote sensing and GIS Techniques: a case study in Nagenahalli watershed, Doddaballapur taluk, Bengaluru rural district, India. Int J Curr Microbiol App Sci 6(8):314–331

Nautiyal S, Bhaskar K, Khan YI (2015) Biodiversity of semiarid landscape: baseline study for understanding the impact of human development on ecosystems. Springer, Berlin

Pandey DN (2002) Traditional knowledge systems for biodiversity conservation. Organization of the United Nations (FAO) Forestry Paper. FAO Rome Italy 22–41

Phillips EA (1959) Methods of vegetation study

Pilgrim SE, Cullen L, Smith D, Pretty J (2007) Hidden harvest or hidden revenue? The effect of economic development pressures on local resource use in a remote region of southeast Sulawesi, Indonesia. Indian J Tradit Knowl 6:150–159

Ramakrishnan PS (2005) Mountain biodiversity, land use dynamics and traditional ecological knowledge. In: Huber UM, Bugmann HKM, Reasoner MA (eds) Global change and mountain regions. Adv Glob Chang Res 23 (Springer, Dordrecht)

Saxena AK, Singh JS (1982) A phytosociological analysis of wood species in forest communities of a part of Kumaun Himalaya. Vegetation 50:3–22

Swamy RN (2017) Protection of traditional knowledge in the present IPR regime: a mirage or a reality. Indian J Public Adm 60(1):35–60

Warren DM (1996) Indigenous knowledge, biodiversity conservation and development. Sustainable development in third world countries. Appl Theor Perspect 81–88

Warren DM, Rajasekaran B (1993) Putting local knowledge to good use. J Agric Environ Int Dev 13(4):8–10

Zent S, Maffi L (2009) Final REPORT on Indicator No. 2: methodology for developing a vitality index of traditional environmental knowledge (VITEK) for the project 'Global Indicators of the Status and Trends of Linguistic Diversity and Traditional Knowledge, Terralingua

Chapter 5
Spatio-Temporal Dynamics
of Rural-Urban Interface and FMV

5.1 Land Use Land Cover Analysis of Rural-Urban Interface of Bengaluru

In recent years, ecosystems have been dramatically disrupted by increasing human activities and their service functions have been seriously compromised. At the same time, potential negative environmental effects due to the intensification of agriculture cannot be ignored (Ramankutty et al. 2018; Rockström et al. 2017; Springmann et al. 2018). To effectively manage natural resources and associated ecosystem services, multi-decade studies provide greater insights into changes in agroecosystem and the underlying ecological and socio-economic drivers than short-term analyzes (Dearing et al. 2012). Multi temporal remote sensing studies provide opportunity to reduce the variability related to shift and disruption of land use land cover change (LULCC) and their effects. Dynamics of Land use and Land cover trend reveals possible factors for change in biophysical landscape and in this case the structures of Field margin Vegetation (FMV). Similarly, urbanization, agricultural intensification, cropping pattern change over a period of time are considered major factors in changing field margin vegetations. The systematic analysis of local scale land use dynamics is conducted over a range of timescales, helps to uncover the driving factors of FMV dynamics.

Land use land cover changes (LULCC) plays very important role in the study of spatial change. Land use/land cover and human/natural modifications have largely resulted in deforestation, biodiversity loss, global warming etc., (Mas et al. 2004). LULCC are directly related to these environmental problems. The driving factors for land use/land cover shift are the rising population, fragmentation of land holdings, socio-cultural changes and increasing socio-economic needs. These factors result in unplanned and uncontrolled changes in LULC. In this research Landsat satellite images are used to identify synoptic data of spatial information of study area. The aim of the study is to recognize change detection of specific classes of land use/land cover over a period of time in the study area (Fig. 5.1).

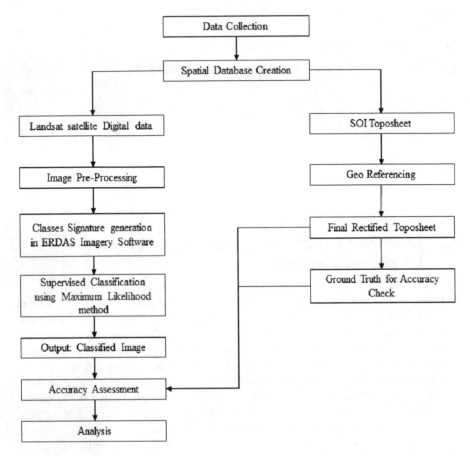

Fig. 5.1 Methodological framework for decadal land use land cover change estimation in the northern transect of Bengaluru

5.2 Methodology and Data Used

Satellite Data

Landast TM and Landsat ETM+ (path 144, row 51) were used in this study. The Landsat 5 image for the dates 12th February 1991 & 07 March 2011 with 30 m spatial resolution and Landsat 7 and Landsat 8 images for the date 03 March 2001 & 10 March 2018 with 30 m spatial resolution were downloaded respectively from Earth Resources Observation and Science (EROS) Center, USGS (https://www.usgs.gov/centers/eros/data). The dates of all images were chosen to be as closely as possible in the same vegetation season. All visible and infrared bands (except the thermal infrared) were included in the analysis. Remote sensing image processing was performed using ERDAS Imagine 14 and ArcMap 10.3.1.

Other tools and data used are Google Earth Pro and standard topographic maps. The training sites and test sites maps, generated from Google Earth Pro interpretation, ground truth data using handheld GPS device and Toposheets for the decadal change were digitized for the goal of creating a spatial database.

Image Classification and Accuracy Assessment

In this study, totally, six LULC classes were established as agriculture, built up, barren land, plantations, water body and scrub land. Landsat images of four different points of time were compared with supervised classification technique. In the supervised classification technique, four-layer stacked images with different dates are independently classified. A Supervised classification method was carried out using training areas and test data for accuracy assessment. Maximum Likelihood Algorithm was employed to detect the land cover types in ERDAS Imagine 14. The accuracies developed by comparing randomly and independently selected test pixels with the ones used in classification.

Accuracy Assessment and Validation of LULC Maps

Error matrix and kappa coefficient-based accuracy assessment.

The confusion matrix-based accuracy assessment and Kappa coefficient were considered for assessing the performance of the classification system (Rogan and Chen 2004). Due to a lack of past ground truth data, it is difficult to carry out accuracy assessment for all the classified LULC maps (except current). Therefore, Google Earth Pro and recent OLI-based accuracy assessment represents the overall accuracy measure for all the classified maps as all the datasets were calibrated on a single platform. 410 random sample points belonging to all the corresponding LULC classes were selected through the stratified sampling method and verified against the reference data. The results showed an overall accuracy of 87.59% and kappa index of agreement value of 0.86. The accuracy levels for agriculture, built up, barren land, plantations, water body and scrub land were 89.84%, 86.63%, 88.48%, 88.30%, 85.95% and 87.35% respectively. All classes were over 90% in terms of class-wise accuracy except the agricultural land. Similar accuracy levels are expected from the past LULC maps with very little deviation. It is evident that the present classification approach has been effective in producing consistent results irrespective of differences in spatial, spectral and radiometric resolution of satellite images from the accuracy assessment.

5.3 LULCC in Rural-Uurban Interface of Bengaluru Over Four Decades

An urban growth can be termed as linear or non-linear depending on the spatial orientation of growth and can be scattered or continuous depending on the distribution and density of population. Fringe growth is extensions near sub-urban areas. Depending

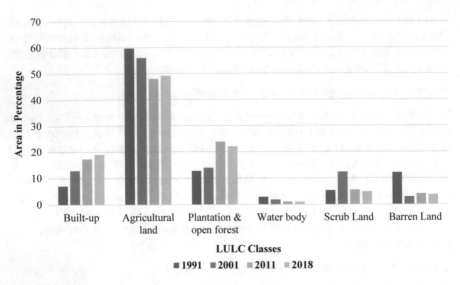

Fig. 5.2 Percentage of area under different land use land cover classes at four points of time in the northern transect of Bengaluru

on the spill out of built-up area with respect to administrative boundary, the growth can be over bound or under bound. Increase in built-up area in rural-urban interface is inevitable and results show that built-up area has increased from 7.10% in 1991 to 19.06% in 2018. The results using different modelling techniques indicated the growth of suburban towns such as Yelahanka, Hesaragatta, Hoskote and Attibele with urban intensification at the core area (Ramachandra et al. 2013). The impact of peri-urbanization around Yelahanka is evident from the results in terms of increased built-up area.

Figures 5.2 and 5.3 show the trend of land use land cover transition in the study landscape from 1991 to 2018. The analysis reveals significant spatio-temporal shift in the landscape. While the increase in the area of plantation/vegetation in the landscape is substantial, i.e. the increase from 12.22% in 1991 to 23.82% in 2011 and 22.03% in 2018, the shift in the dynamics of plantation is seen mainly in the urban area of the transect; while in the rural area of the transect, the increase in the vegetation area is noticeable at present. It may be related to agroforestry practices, in particular *Eucalyptus* plantation, which is still taking place in rural areas of the study area, while in urban and transect transition areas there is a shift from other land-use areas classified as including planting to built-up areas. The increase in built-up patches in the western part of the transition zone can be attributed to the urbanizing impact of the city of Doddaballapur rather than to the impact of the expansion of the city of Bengaluru. The decrease in barren land is also important for the transect, which indicates 12.11% in 1991 and 3.77% in 2018. This can be attributed to planting schemes to restore the ecosystem. There has been almost 10% decrease in agricultural land during time period of 1999 to 2018, which is expected to have an effect on

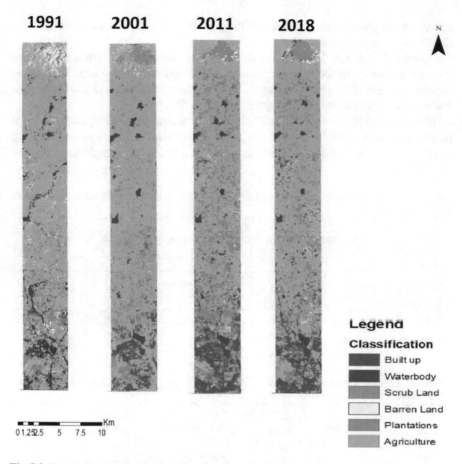

Fig. 5.3 Decadal land use land cover change in the selected transect of northern Bengaluru (Nautiyal et al. 2020)

the system of agricultural production. Changes in built-up area, agricultural land and vegetation cover (specifically planting) are likely to have a possible effect on the area (in terms of increase in green patches) and composition of field margin vegetation. Further studies are needed to correlate land use land cover shift and its reciprocal changes in FMV.

5.4 Interclass Decadal Change of LULC

Land use and land cover change using remote sensing and GIS techniques is very useful in monitoring and managing of natural resources. LULCC gives a trend of change in classes over a period of time. Inter-class changes in the land use and land cover change gives trend of classes shifting from one to another to understand the dynamics of landscape change.

Tables 5.1, 5.2, and 5.3 show the decadal inter-class change matrix of land use land cover classes in the rural-urban transect of northern Bengaluru. The yellow highlighted cells show the percentage of area of a particular class retained the same and other cells shows the percentage of interclass change.

It is observed that from 1991 to 2001, 68.08% agriculture area was retained as it is, whereas, 10.24% and 8.07% of area were converted to Plantations and Built-up respectively. More than half of the barren land during 1990s changed to productive

Table 5.1 Inter-class transition matrix of LULC (1991–2001) in the northern transect of Bengaluru (area in percentage)

Classes	1991-2001					
	Agriculture	Barren	Built up	Plantations	Scrub	Water Body
Agriculture	68.08	3.38	8.07	10.24	8.8	1.43
Barren	45.24	3.45	9.09	10.3	31.4	0.51
Built-up	7.12	2.1	78.36	6.47	4.57	1.38
Plantations	41.99	0.58	7.82	34.4	15.07	0.14
Scrub Land	49.08	0.87	29.67	4.6	11.91	3.86
Waterbody	17.2	0.75	25.57	3.96	10.26	42.25

Source Authors' analysis

Table 5.2 Inter-class transition matrix of LULC (2001–2011) in the northern transect of Bengaluru (area in percentage)

Classes	2001-2011					
	Agriculture	Barren	Built Up	Plantations	Scrub	Waterbody
Agriculture	54.37	3.29	11.07	29.37	1.81	0.07
Barren	75.81	5.61	3.99	14.12	0.4	0.01
Built-up	9.81	1.75	79.96	4.78	3.41	0.29
Plantations	22.95	3.84	8.39	58.81	5.89	0.07
Scrub	24.27	8.89	9.43	49.8	7.44	0.11
Water Body	32.01	0.46	19.3	25.78	3.32	18.82

Table 5.3 Inter-class transition matrix of LULC (2011–2018) in the study transect (area in percentage)

Classes	2011-2018					
	Agriculture	Barren	Built up	Dense Scrub	Plantations	Waterbody
Agriculture	74.6	3.65	7.82	0.79	12.67	0.46
Barren land	51.08	12.69	13.4	5.39	17.22	0.12
Built-up	6.76	2.62	82.55	0.43	6.78	0.85
Plantations	34.44	3.73	9.1	4.85	47.08	0.77
Scrub	27.47	4.65	19.35	17.24	30.09	0.96
Waterbody	10.64	0.03	15.03	0	4.66	67.09

area, where 45.24% of area converted to agriculture and 10.30% converted to plantations. Major area of scrub land and water bodies are converted to built-up area in 2001. 42.25% of water body area remains unchanged and 25.57% and 17.2% of its area was converted to built-up and agriculture land respectively. The multifold growth of the city has occurred in last four decades has been reflected in shrinking water bodies. The retention of area under waterbodies as waterbodies was 42.25% (1991–2001), 18.82% (2001–2011) and which has shown a satisfactory retention of waterbodies (67.09%) in the recent decade (2011–2018).

In 2001–2011 inter-class change (Table 5.2), it is observed that 45.63% of agriculture area has converted to other classes, where farmers adopted agroforestry. Impact of agroforestry on restoration of barren land is also reflected during this decadal change with 90% barren land converted to agriculture and plantations. Similar to the trend of the previous decade, built-up area has increased in 2011 replacing 11.07%, 8.39%, 19.30% of agricultural land, plantations and waterbody of 2001 respectively. 54.37% of agriculture area of 2001 remains unchanged in 2011. 29.37% of area changed to plantations and remaining 16.36% area changed to other classes. The most dynamic LULC classes during this decade were barren land and scrub land where only 5.61% and 7.44% of area of respective classes remained unchanged. Scrub land/vegetations of 2001 was primarily utilized for plantations (49.80%) and agriculture land (24.27%) in 2011. Only 18.82% waterbodies and wetlands of 2001 remained unchanged, the rest seems to be encroached or changed for developmental activities like agriculture (32.01%) and built-up (19.30%) in 2011.

Inter-class (Table 5.3) of 2011–2018 shows that 74.60% of agricultural land from 2011 remained unchanged in 2018. 12.67% of area shifted to plantations and only 13.40% changed to built ups. 51.08% of barren land converted agriculture. 13.40% & 17.22% of barren changed to built ups and plantation respectively. Only 12.69% of barren area remained unchanged. 66.15% of built up in 2011 remained unchanged in 2018. 23.86% built up area changed to agriculture in 2018. 34.44% of plantation area of 2011 shifted towards agriculture in 2018. 47.08% area of plantation remained unchanged in 2018. In 2018, 30.09% area of scrub land shifted to plantations and 27.47% & 19.35% area changed to agriculture and built-up respectively. Major part of water body remained unchanged in 2018 (67.09% of 2011), 15.03% & 10.64% area of waterbodies changed to built-up and agriculture respectively.

Over all, it is observed that the inter-class change between agriculture and plantations is significant. The reason for this may be, farmers shifting from conventional crops like ragi, maize, other millets towards cultivation of grapes, pomegranate, guava etc. *Eucalyptus* plantations rapidly replaced agricultural area between late 1970 and 2010. In 1995 government of Karnataka allotted area for the construction of International Airport near Devanahalli, Bengaluru Rural district, resulted development and shifting of agriculture area to built-up. The trend of raise in built ups remained same in last 2 decades.

5.5   Village-Wise LULC Change

For micro level understanding of the landscape dynamics, the transect divided into three parts namely urban, rural and transition on the basis of population density, distance from the city core and extent of subsistence agriculture and land transformation.

Table 5.4 shows percentage of area under different LULC classes in respective villages. In all the villages, highest percentage of area comes under crop land category. In Muddenahalli village, crop land covers the highest land area which is 47.17%, and barren land, fallow land, settlement and dense vegetation cover 18.35% 23.52%, 7.91% and 2.15% respectively. In Singanayakanahalli, crop land is 41.77%, settlement, fallow land, dense vegetation, plantation, open barren land and water body covered with 23.70%, 5.44%, 5.98%, 6.59%, 16.51% and 0.01% respectively. In Arakere, crop land is 46.55%, settlement, fallow land, dense vegetation, plantation and open barren land covered with 1.66%, 12.35%, 3.52%, 5.41% and 30.52% respectively. In Kundana, crop land accounts for 34.48%, settlement, fallow land, dense vegetation, plantation, open barren land and water body covered with 2.61%, 31.48%, 7.66%, 17.98%, 15.72% and 0.02% respectively. In Heggadihalli, 39.62% is crop land; whereas, settlement, fallow land, dense vegetation, plantation, open barren land and water body covered with 3.58%, 16.50%, 9.86%, 6.92% 22.61% and 0.09% respectively. In Konaghatta, crop land is covered with 44.89%, settlement, fallow land, dense vegetation, plantation, open barren land and water body covered with 2.80%, 31.53%, 4.86%, 3.42%, 11.73% and 0.75% respectively. Being

Table 5.4 Percentage of area under different land use and land cover classes of six villages located in three different zones of northern transect, Bengaluru

LULC classes	Singanayakanahalli*	Arakere*	Kundana**	Konaghatta**	Heggadihalli***	Muddenahalli***
Total area in ha	315.79	471.34	422.35	739.26	273.53	28.83
Water body	0.01	0	0.02	0.75	0.9	0
Settlement	23.7	1.66	2.61	2.8	3.58	7.91
Fallow land	5.44	12.35	24.52	31.53	16.5	23.52
Crop land	41.77	46.55	31.48	44.89	39.62	47.17
Dense vegetation	5.98	3.52	7.66	4.86	9.86	2.15
Plantation	6.59	5.41	17.98	3.42	6.92	0.9
Open barren land	16.51	30.52	15.72	11.73	22.61	18.35

Source Authors' analysis (*urban, **transition, ***rural)

located in rural-urban interface, transition zone villages are more prone to conversion of agricultural land to other land use classes, more specifically built-up land. The results indicate that the villages in transition zone, i.e. Kundana (24.52%) and Konaghatta (31.53%) have high area under fallow land category. The reasons for having large fallow land area would be many and some of them are as follows. The land is not much remunerative and people have job opportunities in Bengaluru which could easily provide them required income. Another reason that the farmers are leaving land uncultivated, that after a gap for few years the land conversion for developmental activities would be easier which could provide high income to the local people. From the viewpoint of climate change, the availability of surface water as well as ground water is poor, entire land use of the area is dependent on the onset of monsoon showers and climate sensitive agriculture is not encouraging farmers to bring all the available lands under cultivation. There may be another reasons for high percentage of fallow land detected in LULC analysis. Hence to understand the perspective in proper scientific manner there is a need for conducting indepth study on the behavioural economics and governance of socio-ecological systems. Figure 5.4a–f present the visualization of land use land cover analysis of all the six villages.

Assessment of FMV Area (Tree Cover)

The high-resolution Worldview 3 multi spectral image provides clear view of field margin vegetation. With the help of Worldview 3 field margin boundaries were delineated for the year 2017. However, for the year 2005, the Google Earth images were used as an unavailability of high-resolution images at that time. Field margins were identified around 355 agricultural fields in the study area, 55–60 fields in each of the six villages. During the selection of plots, the uniform spatial distribution and coverage of crop diversity in the study area were considered.

Figure 5.5 shows the comparison of areas in field margins at two points of time (2005–2017). The research results indicate that there has been decline in total area under FMV during last 12 years. The vegetation area under field margin was 85.40 ha in 2004/2005 and decreased to 76.69 ha in 2018 as estimated from 355 sample plots. In-spite of increase in vegetation cover in the landscape, the field margin area has been shrinking by 10.20% (Fig. 5.5). It is indicative from the results and field survey that field margins have been either converted to crop area, cleared or merged with farm forestry to some extent. It displays an increase of area under FMV in urban part of the transect by 4.6% during the studied 12 years. On the other hand, FMV area is decreasing (16–20%) in both transition and rural zones of the transects. Assessing the vegetation types in field boundary and analyzing area under FMV over 12 years interval provided insights on various factors that may be responsible for the decline of FMV. It is indicative from the results and field survey that field margins have been either converted to crop area of farm forestry to some extent (Fig. 5.6).

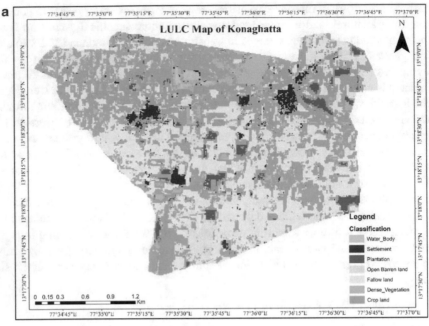

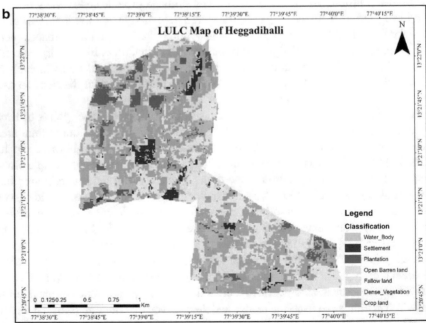

Fig. 5.4 a Land use and land cover of Konaghatta, **b** Land use and land cover of Heggadihalli,
c Land use and land cover of Kundana, **d** Land use and land cover of Arakere (Arekere), **e** Land
use and land cover of Muddenahalli, **f** Land use and land cover of Singanayakanahalli

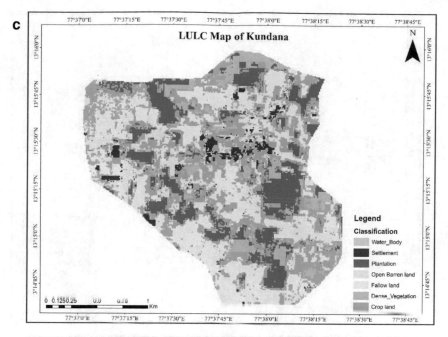

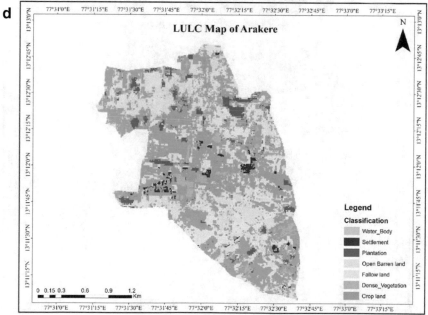

Fig. 5.4 (continued)

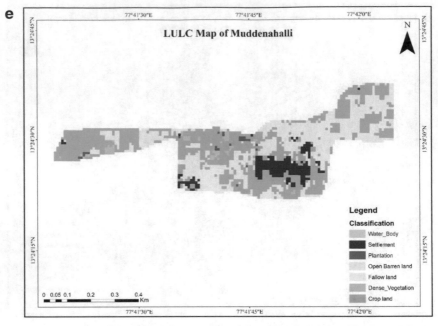

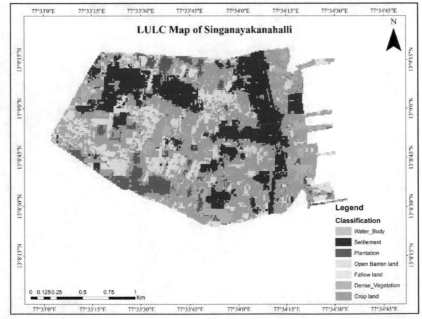

Fig. 5.4 (continued)

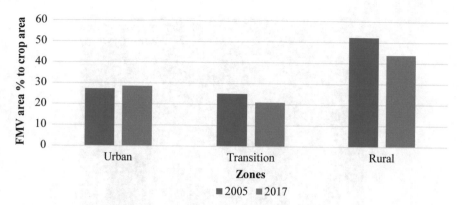

Fig. 5.5 Share of FMV areas to crop areas across the transect (based on the analysis of the sampling plots)

5.6 Vegetation Indices; NDVI and SAVI

The Landsat satellite imagery of medium resolution (30 m) were used to get an understanding of broad vegetation changes over a period of 1991–2018 (1991 2001 2011-TM sensors 2018 OLI TIR sensors) (Table 5.5). The general methodology to assess the vegetation consists of data collection and followed by pre-processing of images for developing vegetation maps. The computation of the vegetation indices, namely, normalized difference vegetation index (NDVI) and the soil-adjusted vegetation index (SAVI) were used for acquiring the information on pixel-wise intensity of vegetation and analysis was carried out following the equation 1 (NDVI) and equation 2 (SAVI). A classification technique was then applied to the NDVI images of 1991 2001 2011 and 2018 using Arc-GIS 10.1 software. NDVI images are obtained by calculating the ratio between the visible Red and near-infrared (NIR) bands of the Landsat satellite image. This emphasizes the characteristics and the different status of plant vigour and species density.

NDVI values are in between −1 and +1, where higher values represent more vigorous and healthy vegetation. According to Gross (2005), very low values (0.1 and lower) correspond to barren areas of rock, sand and snow. Moderate values (0.2–0.3) indicate shrub and grassland, while temperate and tropical rainforests are represented by high NDVI values (0.6–0.8) (El-Gammal et al. 2014).

$$NDVI = \frac{(NIR - Red)}{(NIR + Red)}$$

$$SAVI = \frac{(NIR - R)}{(NIR + R + L)} \times (1 + L)$$

where, NIR = Near Infrared Band
Red (R) = Red Band, L = Correlation Factor (from 0 to 1).

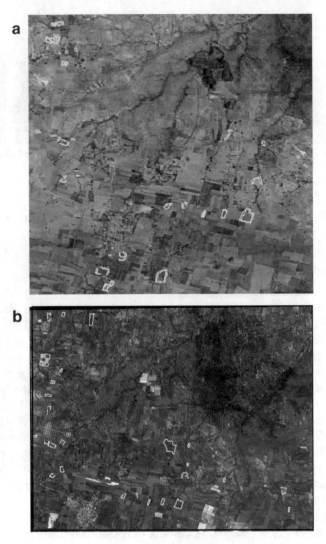

Fig. 5.6 **a** Results of manual digitization of FMV for 2005, **b** Results of manual digitization of FMV for 2007

Table 5.5 Details of satellite data used for vegetation assessment in the northern transect of Bengaluru

S. no.	Sensor	Spatial resolution (m)	Date of data acquisition
1	Landsat—05	30	12-02-1991
2	Landsat—05	30	27-03-2001
3	Landsat—05	30	07-03-2011
4	Landsat—08	30	10-03-2018

The value of the soil factor (L) obtained for SAVI is 0.5 and finally vegetation layers were produced and statistical analysis was carried out to determine percentage of vegetation cover in study region. In addition, detailed analysis of FMV was undertaken in the 13.5% area of the northern transect, Bengaluru. High resolution Worldview 3 image (Digital Globe) captured on 22 October 2017 over the study region was procured for delineating FMV area. The segment area for FMV covers 19250 ha with varying vegetation density. Figure 5.7a and b depicts the result of vegetation mapping from 30 m spatial resolution where vegetation is seen in green. The depicted results obtained for the comparison between Normalized Difference Vegetation Index and Soil-adjusted Vegetation Index using Landsat imagery. The soil factor parameter (L) was taken 0.5 and research shows SAVI index is more sensitive to presence of vegetation when the spatial resolution is medium.

Table 5.6 indicates overall change in vegetation using NDVI and SAVI in the study region. There has been an increase of vegetation area as assessed by both the methods- NDVI and SAVI, 38% and 49% respectively from 1991 to 2018. This would be due to change of main land use to farm forestry; for example, plantation of *Eucalyptus*, coconut and other profitable tree species. India state of forest report (2017) states more than 1000 km^2 increase in vegetation cover which is mostly attributed to expansion of plantation. Both Bengaluru Rural and Bengaluru Urban districts are dominated by agroforestry systems, viz.- agri-silvi-pasture, agri-silvi-horticulture and home gardens (Ashok et al. 2016). The Karnataka State Social Forestry Project was implemented from 1983 to 1984 and 90% of the plantation was covered by planting *Eucalyptus* (Palanna 1996) which has led to deterioration of groundwater table in the area because of those water intensive plantations. As a consequence, the farmers have also adopted extensive plantation of timber and fruit crops instead of traditional agricultural practices, worsening the situation.

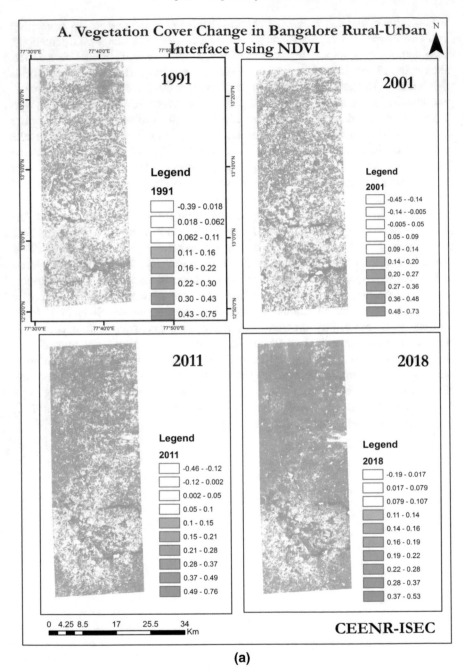

(a)

Fig. 5.7 a and **b** Comparison of NDVI versus SAVI at 30 m spatial resolution for northern transect of Bengaluru

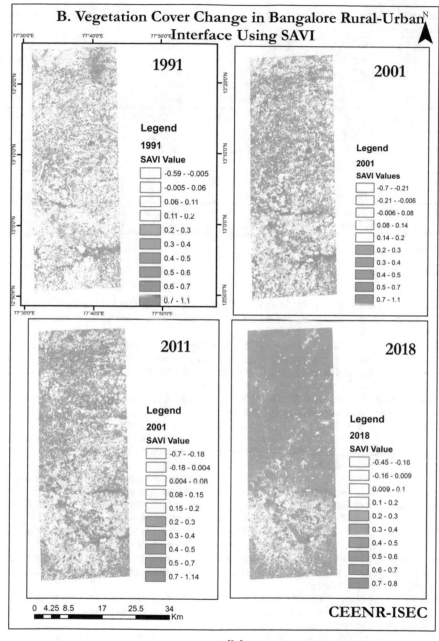

(b)

Fig. 5.7   (continued)

Table 5.6 Analysis of vegetation cover analyzed using NDVI and SAVI for the northern transect of Bengaluru

S. no.	Year	NDVI range (−1 to 1)	SAVI range (−1.5 to 1.5)	Area under vegetation cover (NDVI) (km²)	Area under vegetation cover (SAVI) (km²)	Vegetation area in % (NDVI)	Vegetation area in % (SAVI)
1	1991	0.1 & above	0.2 & above	653.95	440.71	46.20	31.13
2	2001	0.1 & above	0.2 & above	827.61	627.98	58.46	44.36
3	2011	0.1 & above	0.2 & above	812.35	677.16	57.38	47.83
4	2018	0.1 & above	0.2 & above	1192.90	1134.02	84.27	80.11

References

Ashok ST, Vithal N, Ratnakar MS, Shivanand H, Raju C, Halesh GK (2016) Sustainable Silvi based cropping systems for improving socioeconomic status of horticulture farmers—a review. Int J Adv Res Biol Sci 3(7):99–104 (2016)

Dearing JA, Yang X, Dong X, Zhang E, Chen X, Langdon PG, Zhang K, Zhang W, Dawson TP (2012) Extending the timescale and range of ecosystem services through paleoenvironmental analyses, exemplified in the lower Yangtze basin. In: Proceedings of the National Academy of Sciences 109(18):E1111–20

El-Gammal MI, Ali RR, Samra RA (2014) NDVI threshold classification for detecting vegetation cover in Damietta governorate, Egypt. Am Sci 10:8

Mas JF, Velázquez A, Díaz-Gallegos JR, Mayorga-Saucedo R, Alcántara C, Bocco G, Castro R, Fernández T, Pérez-Vega A (2004) Assessing land use/cover changes: a nationwide multidate spatial database for Mexico. Int J Appl Earth Obs Geoinformation 5(4):249–61

Nautiyal S, Goswami M, Nidamanuri RR, Hoffmann EM, Buerkert A (2020) Structure and composition of field margin vegetation in the rural-urban interface of Bengaluru, India: a case study on an unexplored dimension of agroecosystems. Environ Monit Assess 192(8):1–6

Palanna RM (1996) Eucalyptus in India. In: Reports submitted to the regional expert consultation on Eucalyptus, vol 2. RAP Publication: 1996/44, Bangladesh, pp 1–280

Ramachandra TV, Aithal BH (2013) Urbanisation and sprawl in the Tier II City: metrics, dynamics and modelling using spatio-temporal data. Int J Remote Sens 3:66–75

Ramankutty N, Mehrabi Z, Waha K, Jarvis L, Kremen C, Herrero M, Rieseberg LH (2018) Trends in global agricultural land use: implications for environmental health and food security. Annu Rev Plant Biol 69:789–815

Rockström J, Williams J, Daily G, Noble A, Matthews N, Gordon L, Wetterstrand H, DeClerck F, Shah M, Steduto P, de Fraiture C (2017) Sustainable intensification of agriculture for human prosperity and global sustainability. Ambio 46(1):4–17

Rogan J, Chen D (2004) Remote sensing technology for mapping and monitoring land-cover and land-use change. Prog Plann 61(4):301–325

Springmann M, Wiebe K, Mason-D'Croz D, Sulser TB, Rayner M, Scarborough P (2018) Health and nutritional aspects of sustainable diet strategies and their association with environmental impacts: a global modelling analysis with country-level detail. Lancet Planet Health 2(10):e451–61

Chapter 6
Delineation and Monitoring of FMV

6.1 Remote Sensing in Vegetation Study; Prospect for FMV Mapping

It has been established that biodiversity of agroecosystems has immense ecological, social and economic significance. Nevertheless, there is mere recognition for ecological services provided by FMV but a dearth exists for any scientific investigation of spatio-temporal analysis of non-crop plant species present in the Field Margin. More precisely, in this era of extensive use of remote sensing and advanced computation technology in spatio-temporal assessment, field margins have not got any attention which might be because of their complex and heterogenous composition. There are several factors influencing estrangement of remote sensing and GIS technologies from studying FMV; which may include it's not so prominently established functions, perceived prevalence as a scattered and insignificant element in the agroecosystem, linear and slim physical structure making monitoring and delineation a difficult process. Although this element with complex structure and high biodiversity and plays a key role in maintaining the biological equilibrium in agricultural landscapes along with diverse services (De Cauwer et al. 2005).

Manual field studies can help to understand the basic structure of FMV and ameliorate understanding of FMV to socio-ecological sustainability. However, manual studies require huge inputs in terms of resources, time and energy. Therefore, using conventional methods, the delineation of FMV for large landscape is not possible. Thereby, the need of developing an efficient method for delineating and mapping FMV area was realised to map the area and to assess the spatio-temporal changes. Remote sensing data would be more meaningful in this regard, but critical limitation is to segregate FMV from other vegetation due to similarity in spectral reflectance in both FMV and other types of vegetation.

Mapping of vegetation, identification of type and species are important for analysing vegetation dynamics, quantifying spatial patterns and assessing environmental impacts (Suo et al. 2019; Xiao et al. 2004). It is important to obtain status of vegetation cover to initiate conservation and restoration programs (Egbert et al. 2002; He et al. 2005). Traditional and/or conventional mapping methods like field surveys, map interpretations and ancillary data are commonly followed methods; however, these methods for mapping vegetation in bigger scale are time consuming, data lagged and expensive too. Mapping of natural resources using Remote Sensing Technology offers a practical and economical means of study of change in vegetation cover with respect to time and spatial context for large area (Langley et al. 2001). Over the past four decades, remote sensing imagery has been acquired by a range of air-borne and space-borne sensors from panchromatic to multispectral and hyperspectral sensors with wide range in the spectrum. In the past decade, development in space borne image capturing with high spatial resolution, offers the greatest amount of geographical details and useful in many applications like, simulation, precision mapping, assessment etc.

Satellite image classification is a powerful technique to extract information from huge number of satellite images. Various classification algorithms (Supervised and Unsupervised Classification) are generally employed for classifying different levels of classes of features in GIS platform using widely available remote sensing data, based on the requirements. There is a strong need of effective and efficient mechanisms to extract and interpret valuable information from massive satellite images (Abburu and Golla 2015), and image classification techniques are used to do so. Availability of data and the extent of knowledge on the linkages between the remote sensing data and required assessment determine the types of analytical approach (Holloway and Mengersen 2018). Some statistical-based approaches are Maximum Likelihood, Cluster Analysis, and machine learning-based algorithms include Artificial Neural Networks (ANNs), Decision Tree (DT) etc., and Fuzzy ArtMap. Previous study shows Fuzzy ArtMap has been proven to be the most efficient algorithms, followed by the Multi Layer Perceptron (MLP). Another study by Raczko and Zagajewski (2017) compared three machine learning classifiers namely, ANN, DT and Support Vector Machines (SVM) and found that maximum accuracy can be achieved by using ANN. Xie et al. (2019) estimated forest classes using MLP with 89.2% accuracy. Artificial Neural Networks (ANNs) have been useful for decades for the development of image classification algorithms which is based on the training data set (Mahmon and Ya'acob 2014). ANN has been described as efficient in classification for tree analysis (Gomez et al. 2010). These algorithms are per-field pixel based and object-oriented classifiers. Maximum Likelihood (Blaschke 2010) is the most popular tool used by most researchers to classify group of pixels into meaningful classes. Maximum Likelihood is one of the most widely used parametric classifier based on mixture of normal distribution with advantages of robustness, good accuracy and precision (Hennig 2004; Sun et al. 2013; Ahmad and Quegan 2012). Maximum Likelihood often assumes that the distribution of the data within a given class obeys a multivariate Gaussian distribution. Each pixel is assigned to the class with the highest likelihood or labelled as unclassified if the probability values are all below a threshold set (Lillesand et al. 2004).

In the vegetation mapping it is very common that the same vegetation type on ground may have different spectral features in remotely sensed images, whereas, different vegetation types may possess similar spectra (Jiménez and Díaz-Delgado 2015). Thus, it becomes challenging to obtain accurate classification results either using the traditional unsupervised classification or supervised classification. The mentioned classifiers are unable to identify FMV pixels, as the spectral signature would be quite resembling in all types of vegetation. Quantification or mapping of FMV by physical ground survey or manual digitization for each individual farm is hectic and time taking exercise. Therefore, it is essential to work towards development of an alternative and precise method for classification of earth elements which are similar in terms of spectral signature, but independent in terms of social and ecological attributes from other elements with same signature. Most of the classifying algorithms are derived based on the consideration of spectral signature. Object-based classifiers are further modified for efficient classification by integrating structural information such as topology, relative, absolute location etc., with spectral information; GEOBIA (Geographic Object Based Image Analysis) is one such classifier. Image classifiers group the pixels into different classes from the spectral signature registered with the help of training pixels (Blaschke et al. 2014).

A few studies have been conducted worldwide to delineate boundaries of agricultural fields and crop areas using machine learning for image classification algorithms in remote sensing (Turker and Kok 2013; Belgiu and Csillik 2018). Among the recent studies, Masoud et al. (2020) have developed multiple dilation fully convolutional network (MD-FCN) and novel super-resolution semantic contour detection network using transposed convolutional layer in fully convolutional neural networks for detection of field boundaries. Persello et al. (2019) defined the physical edges of field boundaries and used Fully Convolutional Network (FCN) to detect field boundaries in smallholder farms in Africa. They also identified the problems regarding automated delineation of smallholding farm boundaries because of their different size and shapes, non-uniform distribution of fields and mixed cropping patterns. A socioecological study in the selected study area reveals that field boundaries in the rural-urban interface are rich with more than 80 floral species and exhibit a good composition of native and exotic species (Nautiyal et al. 2020). In the existing studies, delineation or mapping of crop land, agricultural field boundaries have been extensively done using both traditional and deep learning techniques, where differentiation of boundary vegetation has not been attempted. Those algorithms using training data set in image classification tools are not able to map vegetations in field boundaries which have complexities regarding both physical attributes (neighbourhoods) and non-separable reflectance from other vegetations.

Selecting suitable variables along with spectral signature is a critical step for successfully implementing an image classification (Lu and Weng 2007). Using high resolution remote sensing data like WorldView 3, it is easy to identify FMV with visual interpretation. Object-based image classification technique integrated with GIS tools are easy and faster way to do systematic work. Like other vegetation, Field Margin Vegetation (FMV) has the same spectral reflectance value; therefore, attempt has been made to develop classification algorithms based on statistical analysis of the pixel histogram of the features in the image that distinguish FMV from other features or vegetation which has been found almost impossible using available classifiers. The chapter discusses three algorithms applied for delineation and mapping of FMV and the accuracy of those results.

6.2 Methodology for Delineating Field Margin Vegetation

Improvisation of methods with multiple algorithms have been designed considering FMVs' physical characteristics, patterns, neighbouring features and spectral reflectance. Preparatory stage of the methodology involves study of structural attributes for FMV in relation with other vegetation types, crops and other land use land cover classes. A database has been generated for FMV through a field survey of randomly selected fields in the study area, collecting ground control points. Following the field survey, land use land cover analysis of the selected area has been done which facilitated the steps for significant physical features associated with FMV and probable errors because of similar reflectance of different features. Further, histogram analysis has been done in this regard to rectify such errors. These exercises have been followed by extensive ground verification. An exhaustive list of constraints based on structural attributes to be applied for defining FMV has been prepared based on visual interpretation of WorldView 3 image and ground truthing. Further exploration of programming platforms has been done to select the appropriate language.

Database used:

  WorldView 3 panchromatic and multispectral image with 0.4 and 1.24 m spatial resolution respectively have been used to delineate the FMV, and other land use and land cover classes. WorldView 3 data with 1.24 m multispectral resolution, short-wave infrared (SWIR) resolution and temporal resolution of one day which was acquired on 22nd October 2017 meets the requirement of land use land cover mapping at a very fine scale (Table 6.1).

  The methodology is divided into three major stages; first, pre-processing stage where data collected and processed; second, algorithm core where actual writing the codes, involves importing of data, classification of FMV and other vegetation using the remote sensing data and running the algorithm in the Julia programming language to get the final output; third is the post processing and accuracy assessment stage (Fig. 6.1). Selection of suitable classification technique and sufficient number of training samples are two major pre-requisites for accurate classification. Experimentation with different classifiers and several non-spectral variables considered

Table 6.1 Description of data used in the study of northern transect, Bengaluru

Sl. No	Sensor	Bands	Spatial resolution (m)	Date of data acquisition
1	WorldView 3	Multispectral band	1.24	22 October 2017
2	WorldView 3	Panchromatic band	0.4	23 October 2017

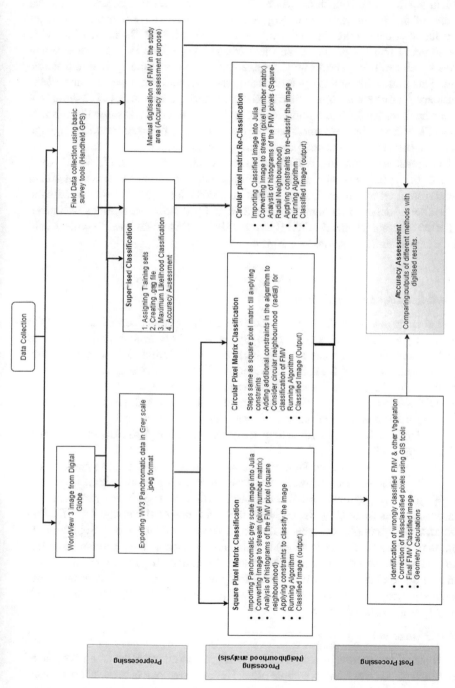

Fig. 6.1 Steps involved in developing accurate methodology for delineating field margin vegetation

for defining different algorithms; where, three algorithms have been selected for the purpose of this target assessment. Rigorous field survey was conducted to characterize the physical features of FMV and their neighbourhoods. During the field visit it is observed that the width of the Field Margin Vegetation is 1–5 m depending on the species of trees, shrubs and herbs, also observed that the agricultural fields, plantations and roads are the neighbouring features/land use types for the FMV. Considering all these physical characteristics, algorithm was scripted in Julia. An attempt has been made in this study in developing an effective classifier algorithm to extract FMV from high resolution satellite image using Julia programming language with basic principles and techniques for image classification.

The pilot area dataset (Fig. 6.2) covers an area of 2084.40 ha with 37,042 × 20,562 pixels of 0.3 × 0.3 m resolution. Panchromatic image of WorldView 3 with 40 cm spatial resolution which had been acquired on 23rd October 2017, provided with the application of default radiometric corrections and geometric corrections was taken up for the analysis. Handheld GPS (Garmin GPSMAP64s) was used to collect field data for the accuracy assessment purpose.

Pre-processing

Exporting Image to .jpeg format

Julia language is a scientific open source programming language that combines the interactivity and syntax of scripting languages as Python, Matlab and R, with the speed of compiled languages such as Fortran and C (Lubin 2015). In this study Julia is chosen as a tool for classifier algorithm because, it is simple to code, as it runs like C, but reads like Python. Julia cannot read geo-referenced image like any other GIS software; therefore, the images are exported to normal.jpeg format without damaging the spatial resolution, then it was exported in grey scale using ArcMap 10.3.1. Algorithm has been written to process grey scale image to ease out the course of image processing.

Supervised classification: Supervised classification is performed to generate the image with defined classes, where it can be used as input for the algorithm. Classified image is used as input for algorithm 3 only, in the other two algorithms, panchromatic image is used as input (Fig. 3a, b). With the assistance of the image classification toolbar, training samples are created for the classes we want to extract. Signature file from the training data set is developed, which is then used by the multivariate classification tools to classify the image. Maximum Likelihood algorithm was employed to detect the land cover types in ERDAS Imagine 14. The accuracy assessment has been performed by comparing randomly and independently selected test pixels with the ones used in classification.

FMV Digitization

As a conventional method, field surveying is adopted to quantify the area coverage of FMV in the fields considering sample plots. Here an attempt has been made to quantify area under FMV using Remote Sensing and GIS techniques. Using visual interpretation method, Field Margin Vegetation have been digitised using high resolution

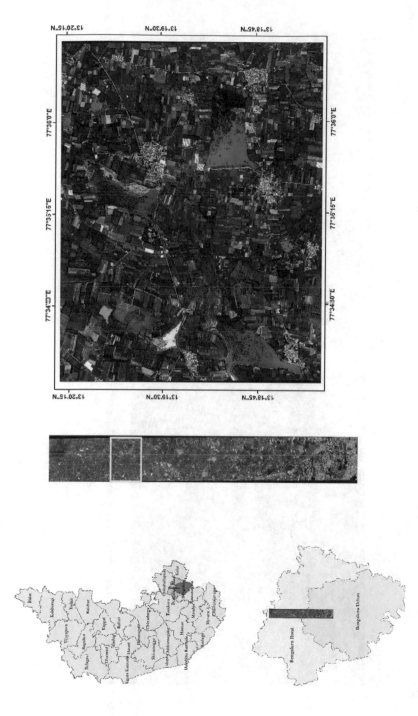

Fig. 6.2 Location and satellite image of Pilot study site located in northern Bengaluru Location and satellite image of pilot study site located in northern Bengaluru

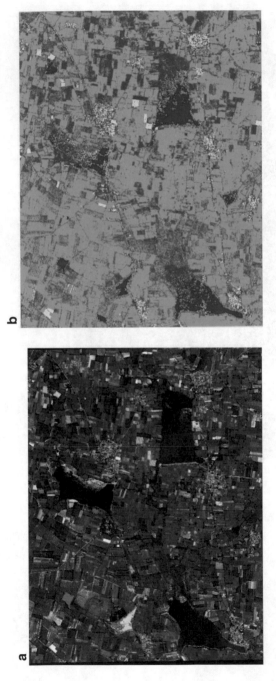

Fig. 6.3 **a** Input for algorithm 1 and 2 (panchromatic image). **b** Input for algorithm 3 (classified image)

satellite image WorldView 3 as a first approach for quantification, as classification algorithms in GIS software packages such as MLP were unsuccessful.

Processing

FMV reclassification using Julia Programming Language: Julia is a flexible dynamic language, appropriate for scientific and numerical computing, with performance comparable to traditional statistically-typed languages. FMV is the vegetation in the border of the fields or vegetation which separates the crop field from any other land use land cover; where, neighbouring features to be considered to identify and map the FMV. It is found that considering these the physical characteristics of FMV, using Artificial Intelligence and Machine Learning Techniques this is possible. Following are the different algorithms tested to delineate the FMV from Worldview 3 data (Panchromatic and Multispectral).

Algorithms: The algorithm 1 is developed to delineate FMV from high resolution satellite imagery without any image processing (raw data) using Julia language. Algorithm 2 is the improved version of algorithm 1 which is developed with processing time efficiency and improved accuracy of the output. Input data used here is the same satellite image used in algorithm 1. Based on previous algorithms trials, it is found that there is scope to increase accuracy and make it time efficient. Thereby, algorithm 3 is developed which is slightly improved version of algorithm 2. The input data used here is classified image. The physical attributes applied as constraints/ conditions in developing the algorithms are based on structural features such as neighbourhoods-crop land, waterbody and built-up land, vegetation density and shadow (Table 6.2).

Algorithm 1 The exported image from Arc Map is imported to Julia. The image has been converted into stream of pixel numbers (pixel number matrix). The neighbourhood of the FMV pixel has been analysed by plotting the histogram of the pixel values. Histograms of the pixel-matrix have been analysed to find the range of pixel values under vegetation and other features. The FMV has same digital value as other vegetation, however, the neighbouring features have difference in the histogram composition; thus, the histogram is analysed to set the conditions on the neighbourhood, so that the algorithm classify FMV precisely. In the general classification algorithm, the classification is performed in individual pixel. But, in this algorithm the command is set in Julia to consider a 13×13 pixel matrix to check if the FMV conditions apply. For this, neighbouring pixels are counted in three categories, viz.-black, grey and white. Further segregation of classes has been done on the basis of concentrations of different categories. To classify Field Margin Vegetation, the conditions areas signed to verify in the matrix if the pixel numbers are less than 5%.

Algorithm 2 Analysis of output from algorithm 1 shows that some of the aquatic vegetation, plantations in agricultural area and scrublands are wrongly classified as FMV. To reduce the error, changes have been made to the earlier algorithm; where, a 40-pixel radial neighbourhood is considered before considering 13×13 neighbourhood matrix. The conditions and rest of the steps are same as algorithm 1. Some changes in the conditions including definition of neighbourhood in terms of crop and built-up pixels have been done.

Table 6.2 Data and parametric description of three algorithms developed to delineate the FMV in northern transect of Bengaluru

Algorithm	Algorithm 1	Algorithm 2	Algorithm 3
Satellite image	WorldView 3 panchromatic	WorldView 3 panchromatic	WorldView 3 multispectral image
Spatial resolution (m)	0.3	0.3	1.24
Pre processing (image format)	Exporting image in grey scale	Exporting image in grey scale	Maximum liklihood classification and exporting image in grey scale
Programming language	Julia	Julia	Julia
Histogram analysis	0–255 pixel range analysis	0–255 pixel range analysis	4 classes pixel analysis
Pixel matrix	Square pixel matrix	Square + radial pixel matrix	Square + radial pixel matrix
Pixel matrix	13 × 13 pixel matrix	40 pixel radial neighbourhood	40 pixel radial neighbourhood
FMV classification	Pixel processing considering neighbourhood features	Pixel processing considering neighbourhood features	Pixel Processing considering Neighbourhood features
Post processing	Georeferencing and raster to vector conversion	Georeferencing and Raster to vector conversion	Georeferencing and Raster to vector conversion

Algorithm 3 Following the development of algorithms 1 and 2 to delineate FMV, it has been marked that reclassifying FMV is simple and easier from the image with defined classes compared to raw satellite image; where in raw image histogram, more pixel ranges have to be analysed. In this algorithm, instead of Worldview 3 panchromatic image, pre-classified image is used. Using Worldview 3 multispectral image classified image is prepared by applying Maximum Likelihood method. The conditions of Algorithm 2 are used in this method with specific changes and refinement in the definition of neighbourhood pixels in the radial matrix. Conditions are enhanced to classify vegetation and command was to change vegetation pixels on square neighbourhood to FMV with refined definitions.

Histogram analysis

Histograms as depicted in Fig. 4a, b, are graphs showing the number of pixels in panchromatic image and classified image (exported to grey scale) as per different intensity values. As both the images are exported into greyscale images, there are 256 (0–255) possible intensities and so the histogram graphically displays 256 numbers showing the distribution of pixels amongst those grayscale values. The histogram of the two images have been analysed further to know the pixel value ranges for different classes in the images. Once the pixels ranges were understood, the values were used for the fixing up the criteria for the FMV classification algorithm. Figure 6.4b shows

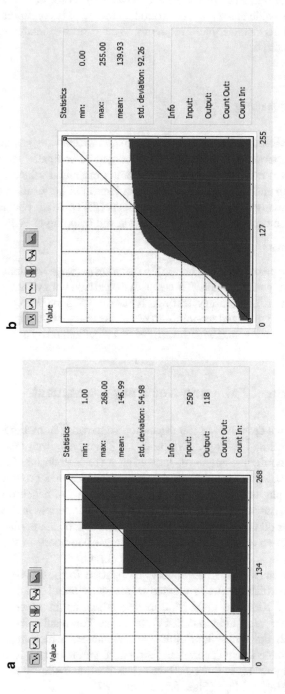

Fig. 6.4   a Histogram for classified multispectral image. **b** Histogram for panchromatic image

the continuous pixel frequency distribution as raw image is considered as input, the image has every possible classes in the ground so the trend is continuous; whereas, Figure 6.4a represents discrete pixel frequency as the classified image has only four pixel values present in it.

6.2.1 Post Processing

The output acquired from Julia was in.jpeg format and without any geo co-ordinates. The image was imported to the ArcMap and the geo co-ordinates are assigned to the image using Geo referencing tool. The wrongly classified vegetation pixels in scrub land, built-up and water bodies have been identified by visual interpretation and corrected manually using mask tool in the ArcGIS software package. Further the final output is considered for the area analysis and mapping of FMV.

Accuracy Assessment

As this is a noble approach, there is no tool in software packages or standard method to assess the accuracy of the results. Therefore, comparison with visually interpreted images is the option for accuracy testing. A total of 573 FMV plots have been randomly chosen and digitized manually to check accuracy of the algorithms. Area of digitized FMV are compared with the area of FMV delineated by three algorithms.

6.3 Delineation of FMV and Accuracy Assessment

The three algorithms described in the three-step framework in methodology using remote sensing, GIS and Julia programming language have been tested for a tile of 2084.40 ha in northern Bengaluru, located at a rural-urban interface. The first step of the methodological framework, i.e. pre-processing has provided the input for the processing phase; in algorithm 1 and 2, .jpeg format of panchromatic image, and in algorithm 3, output of Maximum Likelihood classification. In this study, four land use land cover classes are established viz.-crop land, built-up, plantations, and water body have been established using Maximum Likelihood. In the supervised classification technique, layer stacked images are classified.

It is found that maximum Likelihood classification in the ArcGIS Package is successful to distinguish the major classes in the study area but failed to classify between other vegetation and Field Margin Vegetation, as the spectral reflectance value of forest/other vegetation and FMV are same. The results show that all the performance of all the three algorithms are satisfactory which apparently indicates appropriate selection of spectral and structural variables and definition of the constraints. The algorithms of three different classification methods resulted in three versions of Field Margin Maps (Figs. 6.5, 6.6 and 6.7).

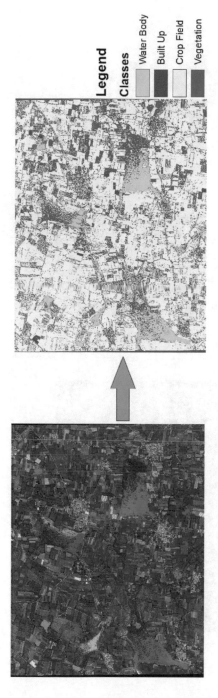

Fig. 6.5 Result of supervised maximum likelihood classification (MLC) for algorithm 3

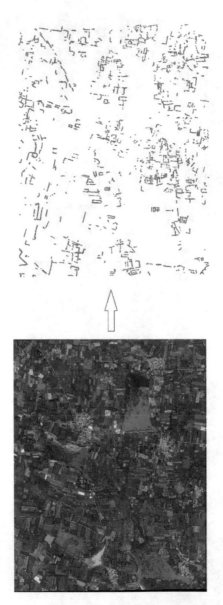

Fig. 6.6 Result of manual digitization of FMV

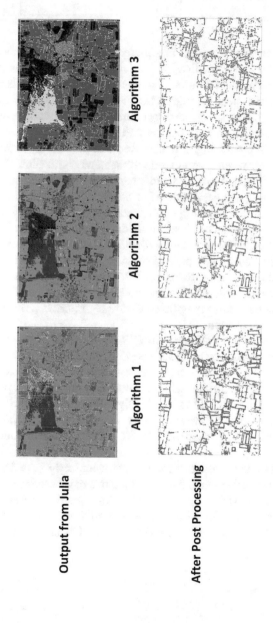

Fig. 6.7 Output images for step 2 and 3 of the methodological framework for all the three algorithms

Table 6.3 Results of accuracy assessment for all the three algorithms

FMV	Manual digitization	Algorithm 1	Algorithm 2	Algorithm 3
Accuracy in %	100	79	67	87
Area in Ha	44.15	34.88	29.58	38.41

The first version of algorithm, where square matrix method has been applied, apparently results in adecent classification by delineating FMV from other classes. Nevertheless, it is observed that non-FMV pixels from plantations, other vegetation and water body have also been wrongly classified as FMV to some extent which is due to the spectral similarity of corresponding vegetation communities. These errors are identified by visual interpretation and rectification has been done using tools in GIS. After post processing, it is observed that the first algorithm classified FMV with 79% accuracy when it compared with manual digitization. To reduce the wrong classification of vegetation as FMV in other neighbouring features except crop fields 2nd algorithm has been designed. After analysing output, it is observed that the algorithm is successful in reducing wrong classification of FMV which included non-FMV pixels, but also observed that reduction in precision with 67% after post processing. The third algorithm is tested with slight modification (as explained the methodology) and using classified image as input, it is observed that the algorithm-3 is successful in reducing the error with finer details and decent precision in classifying FMV with 87%, when it is compared to manually digitized image (Table 6.3). It is also observed that the post processing time of third algorithm is remarkably less as compared to other two algorithms.

Using image classification techniques in GIS platform, the percentage of FMV area coverage in agricultural field has been assessed. The three algorithms have been applied for two tiles of the study transect for assessing the area covered by FMV. The results show that the third algorithm with best accuracy has assessed percentage of field margin vegetation in crop field as 7.49%, algorithm 2 and algorithm 3 assessed share of FMV in crop field as 6.81% and 3.33% respectively (Table 6.4). Efficiency in terms of postprocessing time and error rectification has also been found better for the third algorithm.

Assessment of FMV is an essential aspect that would help in understanding its role in social and ecological systems from micro to macro perspectives. So far FMV aspect of landscape research has not got much attention; therefore, not studied in India as done in Europe and North America. There are traditional methods to study this component in detail (Nautiyal et al. 2020; Goswami and Nautiyal 2020). However,

Table 6.4 Accuracy assessment of FMV for three different algorithms

Algorithm	FMV area in Ha	% of FMV area in crop field
Algorithm 1	96.65	6.81
Algorithm 2	45.56	3.33
Algorithm 3	106.97	7.49

due to variety of reasons (resource, time requirement etc.) the traditional methods are not able to cover the larger landscapes for delineation of FMV. The methods developed with the help of high-resolution satellite data and applying Julia programming are highly scientific and has the prospect to cover larger landscape for assessing and mapping of FMV. This method is scalable to entire northern rural-urban interface of Bengaluru because of two opportunities; at first place, the criteria set for developing algorithms have extensively considered the physical characteristics of the agro-ecosystems of the region, and as a second opportunity, the training data set developed for this methodology is representative of the entire landscape.

India has a commitment for climate change mitigation through creating of additional carbon sink of 2.5–3 billion tonnes of CO_2 equivalents by 2030. The Green India Mission has also stated that 50% of CO_2 sink would be achieved by conservation of forested landscapes and agroforestry systems. Significant amounts of vegetative carbon can also be stored in agroforestry systems or other perennial plantings on agricultural lands (Albrecht and Kandji 2003). Field margins have the potential to help mitigate greenhouse gas emissions from agricultural activities through carbon sequestration in the woody biomass of trees and shrubs as well as in the soil (Thiel et al. 2015). India has approximately 159–160 Million Ha of agricultural land, out of which, almost 110 Million Ha is historically supported directly or indirectly by FMV in traditional socio-ecological system.

Unavailability of an efficient and precise method for mapping and assessment of FMV for enhancing agricultural and ecological benefits was evident for India, which is essential to provide actionable suggestions to policy makers, conservationists and farmers. The study concludes that the described methodology and findings as successful attempt to develop a scientific and robust method with classification algorithms and to overcome the critical limitations in this regard, specifically the similarity in spectral reflectance of FMV with other vegetations. There is distinct need for assessment of field margin vegetations to formulate conservation measures where the method described in this chapter has the potential to scale-up to larger landscape for substantial contribution towards policy formulation in the domain of agroforestry and climate change mitigation. This is a preliminary research in this endeavour and thus has the potential to attract the attention of research community to investigate this unexplored dimension of landscape research for evaluating its role in contemporary issues including sustainable socio-ecological development and climate change.

References

Abburu S, Golla SB (2015) Satellite image classification methods and techniques: a review. Int J Comput Appl 119(8)

Ahmad A, Quegan S (2012) Analysis of maximum likelihood classification technique on Landsat 5 TM satellite data of tropical land covers. In: 2012 IEEE international conference on control system, computing and engineering, IEEE: 280–285

Albrecht A, Kandji ST (2003) Carbon sequestration in tropical agroforestry systems. Agric Ecosyst
   Environ 99(1–3):15–27
Belgiu M, Csillik O (2018) Sentinel-2 cropland mapping using pixel-based and object-based time-
   weighted dynamic time warping analysis. Remote Sens Environ 204:509–523
Blaschke T (2010) Object based image analysis for remote sensing. ISPRS J Photogramm Remote
   Sens 65(1):2–16
Blaschke T, Hay GJ, Kelly M, Lang S, Hofmann P, Addink E, Feitosa RQ, Van der Meer F, Van der
   Werff H, Van Coillie F, Tiede D (2014) Geographic object-based image analysis–towards a new
   paradigm. ISPRS J Photogramm Remote Sens 87:180–91
De-Cauwer B, Reheul D, Nijs I, Milbau A (2006) Dry matter yield and herbage quality of field
   margin vegetation as a function of vegetation development and management regime. J Life Sci
   54(1):37–60
Egbert SL, Park S, Price KP, Lee RY, Wu J, Nellis MD (2002) Using conservation reserve program
   maps derived from satellite imagery to characterize landscape structure. Comput Electron Agric
   37(1–3):141–56
Gomez C, Mangeas M, Petit M, Corbane C, Hamon P, Hamon S, De Kochko A, Le Pierres D,
   Poncet V, Despinoy M (2010) Use of high-resolution satellite imagery in an integrated model to
   predict the distribution of shade coffee tree hybrid zones. Remote Sens Environ 114(11):2731–44
Goswami M, Nautiyal S (2020) Transitional peri-urban landscape and use of natural resource for
   livelihoods. InSocio-economic and eco-biological dimensions in resource use and conservation.
   Springer, Cham, pp 435–457
He ZL, Yang XE, Stoffella PJ (2005) Trace elements in agroecosystems and impacts on the
   environment. J Trace ElemTs Med Biol 19(2–3):125–40
Hennig C (2004) Breakdown points for maximum likelihood estimators of location–scale mixtures.
   Ann Stat 32(4):1313–1340
Holloway J, Mengersen K (2018) Statistical machine learning methods and remote sensing for
   sustainable development goals: a review. Remote Sens 10(9):1365
Jiménez M, Díaz-Delgado R (2015) Towards a standard plant species spectral library protocol for
   vegetation mapping: a case study in the shrubland of Doñana National Park. ISPRS Int J Geo-Inf
   4(4):2472–95
Langley, Kathleen S, Cheshire HM, Humes KS (2001) A comparison of single date and multitem-
   poral satellite image classifications in a semi-arid grassland. J Arid Environ 49(2):401–411
Lillesand TM, Kiefer RW, Chipman JW (2004) Remote sensing and image interpretation
Lu D, Weng Q (2007) A survey of image classification methods and techniques for improving
   classification performance. Int J Remote Sens 28(5):823–70
Lubin M (2015) Dunning I. Computing in operations research using Julia. INFORMS J Comput
   27(2):238–48
Mahmon NA, Ya'acob N (2014) A review on classification of satellite image using Artificial Neural
   Network (ANN). In: 2014 IEEE 5th Control and System Graduate Research Colloquium, IEEE:
   153–157
Masoud KM, Persello C, Tolpekin VA (2020) Delineation of agricultural field boundaries from
   sentinel-2 images using a novel super-resolution contour detector based on fully convolutional
   networks. Remote Sens 12(1):59
Nautiyal S, Goswami M, Nidamanuri, Nidamanuri RR, Hoffmann EM, Buerkert A (2020) Structure
   and composition of field margin vegetation in the rural-urban interface of Bengaluru, India: a case
   study on an unexplored dimension of agroecosystems. Environ Monit Assess 192:520. https://
   doi.org/10.1007/s10661-020-08428-6
Persello C, Tolpekin VA, Bergado JR, de By RA (2019) Delineation of agricultural fields in
   smallholder farms from satellite images using fully convolutional networks and combinatorial
   grouping. Remote Sens Environ 231:111253
Raczko E, Zagajewski B (2017) Comparison of support vector machine, random forest and neural
   network classifiers for tree species classification on airborne hyperspectral APEX images. Eur J
   Remote Sens 50(1):144–54

Sun J, Yang J, Zhang C, Yun W, Qu J (2013) Automatic remotely sensed image classification in a grid environment based on the maximum likelihood method. Math Comput Model 58(3–4):573–581

Suo C, McGovern E, Gilmer A (2019) Coastal dune vegetation mapping using a multispectral sensor mounted on an UAS. Remote Sens 11(15):1814

Thiel B, Smukler SM, Krzic M, Gergel S, Terpsma C (2015) Using hedgerow biodiversity to enhance the carbon storage of farmland in the Fraser River delta of British Columbia. Int Soil Water Conserv Res 70(4):247–256

Turker M, Kok EH (2013) Field-based sub-boundary extraction from remote sensing imagery using perceptual grouping. ISPRS J Photogramm Remote Sens 79:106–21

Xiao X, Zhang Q, Braswell B, Urbanski S, Boles S, Wofsy S, Moore III B, Ojima D (2004) Modeling gross primary production of temperate deciduous broadleaf forest using satellite images and climate data. Remote Sens Environ 91(2):256–70

Xie W, Jiang T, Li Y, Jia X, Lei J (2019) Structure tensor and guided filtering-based algorithm for hyperspectral anomaly detection. IEEE Trans Geosci Remote Sens 57(7):4218–4230

Chapter 7
Overview of a Few Important FMV Species and Crop Influencing FMVs of Rural–Urban Interface of Bengaluru

7.1 *Pongamia pinnata*

Pongamia is a multipurpose tree containing non-edible oil and high percentage of plant nutrients in flowers and seeds. The seeds contain pongam oil 27–36% by weight, which is used for tanning leather, soap, as a liniment to treat scabies, herpes, and rheumatism and as an illuminating oil (Burkill 1966). The oil has a high content of triglycerides, and its disagreeable taste and odor are due to bitter flavonoid constituents, pongamin and karanjin (Duke 1983). The seeds, flowers, leaves and wide range of usage such as minor timber usage, poultry feed, fertilizer, insect repellent, shade tree and windbreak. *Pongamia* seed cakes are rich in plant nutrients, particularly, 4.28% nitrogen and 0.19% sulphur, and critically deficient nutrients like Boron and Zinc (Osman et al. 2009). Studies show significant increase in *Pongamia* production.

In the study transect, *Pongamia* has been found present in sampled quadrats of five villages (Table 7.1). In Arekere village which belongs to urban zone of the transect, there is no record of *Pongamia* tree in our sampled plots. Kundana village in the transition zone observes highest density of *Pongamia* with 88.88 individuals per hectare, whereas the relative density is highest in Muddenahalli village among all the studied villages.

Relative frequency is also highest in Muddenahalli village (14.89) as compared to other villages which indicates better occurrence of the species across the village than others. Kundana village, despite of having high density, relative frequency is comparatively low, shows concentrated occurrence of the tree in a few farm boundaries.

The socioecological household survey reveals the pattern of use of different parts (flower, leaves, seeds) of *Pongamia* and is presented in Fig. 7.1. Respondents reported that the farm households get benefits from *Pongamia* trees on field boundaries in the form of minor timber products, fertilizers from the seeds and seed cakes, and oil extracted from seeds. The oil extracted is used for illuminating diyas (earthen lamp).

© The Author(s), under exclusive license to Springer Nature Switzerland AG 2021
S. Nautiyal et al., *Field Margin Vegetation and Socio-Ecological Environment*,
Environmental Science and Engineering,
https://doi.org/10.1007/978-3-030-69201-8_7

115

Table 7.1 Density and frequency of *Pongamia pinnata* across the transect

Villages	Density (individuals/ ha)	Relative density	Relative frequency
Muddenahalli	55	22.22	14.89
Hegadihalli	37.5	7.58	13.46
Kundana	88.88	8.60	10.95
Konaghatta	7.5	1.49	3.85
Arekere	0	0	0
Singanaykanahalli	30	8.22	9.46

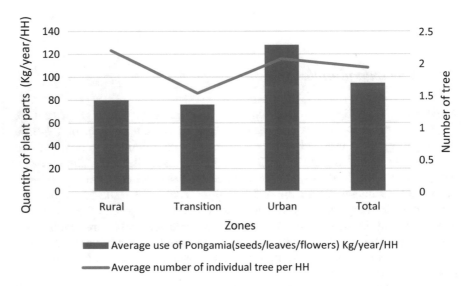

Fig. 7.1 Use of *Pongamia* parts for various purposes in different zones of northern transect of Bengaluru

The households with good number (>5 trees) of trees store the seeds and get the oil extracted. In urban zone, the use of *Pongamia* as fertilizer in their agricultural field is highest with 128 kg per year per household. They use it in the form of pressed cake, powdered seeds, leaf litter etc.

7.2   *Eucalyptus* spp. in Field Boundaries

The vegetation cover in north transect was found to be increasing and the results indicate that there is an increase of 43% in total vegetation cover. Our empirical field information and visual observation revealed that at various location of north transect it was found that the area under vegetation is increased by conversion of agriculture land to farm forestry that includes large scale plantation of eucalypts and many other tree species with economic importance. Some of the forests and barren lands have also been converted into man-made plantations of *Eucalyptus, Acacia auriculiformis* and minor forest species such as Tamarind (Nautiyal et al. 2020). *Eucalyptus* plantation in semi-urban and rural Bengaluru started back in 1930s. Bengaluru North Taluk has 18% of its total geographical area under *Eucalyptus* plantation (Harishkumara 2017). There has been a decrease in *Eucalyptus* plantation because of government policies discouraging the species for its enormous adverse impacts on natural ecosystem and ground water.

In highly intervened field boundaries, *Eucalyptus* spp. are the most common species throughout the transect (Photo 7.1). Although *Eucalyptus* plantation was started through government's policy initiatives during 1930s along with *Casuarina*. With the realization of adverse ecological and hydrological impacts of *Eucalyptus* plantation, despite of its high and casy cconomic return, Government of Karnataka has banned further plantation of *Eucalyptus* spp. in the state. Field observations reveal that new *Eucalyptus* plantations in private land are still carried out by individual farmers.

Parameters representing dominance of *Eucalyptus*, i.e. Relative Density and Relative Frequency in the landscape are tabulated above (Fig. 7.2). The results apparently follow a gradient in sync with the gradient of urbanization, where rural area has the highest values for both Relative Density and Relative Frequency, urban zone has the lowest values for the same. The basal area scores for the three zones are 51.29 cubic feet per acre, 37.36 cubic feet per acre and 4.56 cubic feet per acre along the increasing gradient of urbanization in the transect.

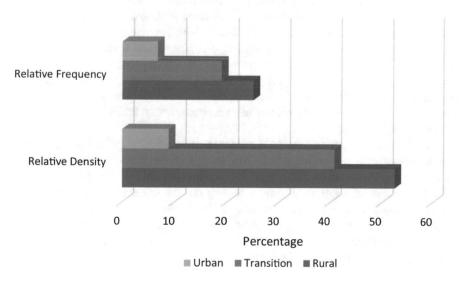

Fig. 7.2 Variables describing the relative dominance of *Eucalyptus* spp. in three zones of northern transect of Bengaluru

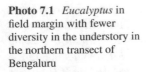

Photo 7.1 *Eucalyptus* in field margin with fewer diversity in the understory in the northern transect of Bengaluru

Effective afforestation started post-independence in the division and thereby farm forestry with short-term economic benefits got more attention. Bengaluru Forest Division (that covers the study area) is one of the few sites where a number of exotic *Eucalyptus* spp. have been introduced from various parts of the world. *Eucalyptus* species like *Eucalyptus bicolor*, *Eucalyptus robusta*, *Eucalyptus citriodora* etc., have been raised in both government and private lands. *Eucalyptus* plantations have been done through two methods, viz.-pit and trench mound, where the *Eucalyptus* Working Circle of the Division covers more than 52,720 acres of land (17.2 3% of the area under forest department) and about 15,000 acres available for plantation (based on Working Plan, Bengaluru Forest Division). The first crop of *Eucalyptus* takes 10 years to harvest and the consequent coppice crops take 8 years. These efforts of Forest Division have diffused to private lands which is very evident from the significant dominance of *Eucalyptus* spp. in field boundaries. Data shows that only in Kundana Gram Panchayat where one of the study village is located, in two years, area under *Eucalyptus* plantation has been increased from 804.95 ha (2012–13) to 1179.56 ha (2014–15).

Water stress and nutrients limitations have been stated as the major limiting factors for the growth of *Eucalyptus* spp. in the Dry zone Southern India. If these two stresses are managed, the growth in first year typically increases fivefold (Calder et al. 1993). The high intake of water and nutrients is very much evident in the form of impact on other vegetations in field margins where *Eucalyptus* are planted (Photo 2.1). As stated in (https://www.fao.org/), Australian studies (Hamilton 1964, 1965, Davidson 1967) and South African study (Scott 1991) showed soils under eucalypts were more water repellent than adjacent similar soil.

7.3 *Vitis* **spp.**

Bengaluru Urban and Bengaluru Rural districts come under the most prominent grape (*Vitis vinifera, Vitis labrusca*) growing districts of the State. Karnataka has doubled the area under grape cultivation in seven years, to 20,400 hectares in 2013–14 from 9700 hectares in 2007–08. In Karnataka, 15 districts have been included in National Horticulture Mission since 2005–06; Bengaluru Rural and Bengaluru Urban districts are two among those. The crops covered under this mission are mango, grapes, pomegranate, banana, pineapple, cashew, cocoa, pepper, ginger, aromatic plants and flowers. Grape has been one of the rapidly adopted crops by the farmers of the study area. Grape production in Nandi Valley (which covers the study transect) is a traditional livelihood which has got tremendous boost from promotion of wineries through government policy. Government encouragement of wineries in Nandi valley has been a key driver in this regard. Area under grape cultivation was increased by 222% from 1998–99 (4330 ha) to 2008–09 (13950 ha) in Bengaluru Rural district (Department of Horticulture, Government of Karnataka).

Grapevines require water in abundance and on time. Water stress induces reduced inflorescence initiation and shoot growth (Sahoo et al. 2009). On the other hand, direct sunlight exposed grape berries go through fast ripening and products are of good quality. In grapevine, developing water stress reduced inflorescence initiation in conjunction with reduced shoot growth (Kliewer and Lider 1968). Canopy management in grape cultivation is an imperative practice optimize sunlight interception, photosynthetic capacity and fruit micro-climate to improve fruit grape and wine quality (Smart et al. 1990; Ray and Choudhury 2015). To fulfil this, tree providing shades surrounding the cultivation and canopies are removed. The development of environment for quality grape production results in change in other characteristics of the agroecosystem by affecting the water share of other crops and removal of vegetation from field boundaries. More than 80% of the grape cultivation is confined to the hot semi-arid agro-climatic region of India. As the growth of the grape vineyards in semi-arid tropics is high, the problems of irrigation water availability and salinity are getting aggravated. Degradation of soil health due to use of indiscriminate application of fertilizers, soil drenching of pesticides and change in irrigation methods are prevalent in those areas (Ray and Choudhury 2015).

Although, the impacts of grape cultivation on soil and water managements have not been reported by the farmers, the removal of trees from the field boundaries has been evident in the study area (Fig. 7.3). The results show that the grape fields in the transition zone do not have a single tree on the field boundaries (Photo 7.2). Thus, the changing cultivation practices and crops are expected to jeopardize the natural and semi-natural field boundary vegetations if required attention is not given on time.

Market dependence for production and for consumption was closely integrated among households indicating the market involvement in agroecosystem. Agricultural practices around Bengaluru, is market driven. Horticulture department of Karnataka has recognised the grape growing districts like Bengaluru (Urban & Rural), Chikkaballapur, Kolar, Bagalkot and Belgaum and has established Karnataka Wine Board

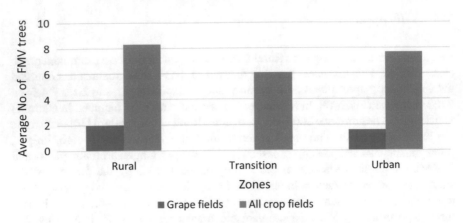

Fig. 7.3 Presence of FMV tree per plot of grape field versus all crop fields

Photo 7.2 The boundaries
of grape fields with no
FMV in the
northern transition of
Bengaluru

in 2007 to help in the process of establishment and monitoring of wineries. This is
encouraging farmers in and around Bengaluru (Urban & Rural) to grow grapes.

References

Burkill JH (1966) A dictionary of economic products of the Malay Peninsula. Art Printing Works,
    Kuala Lumpur, vol 2
Kliewer W M Lider LA (1968) Influence of cluster exposure to the sun on the composition of
    Thompson Seedless fruit. Am J Enol Vitic 19(3):175-184
Ray P, Chowdhury S (2015) Popularizing grape cultivation and wine production in india-challenges
    and opportunities. Int J Soc Res 4(1):9–28
Smart RE, Dick JK, Gravett IM, Fisher BM (1990) Canopy management to improve grape yield
    and wine quality-principles and practices. South Afr J Enol Vitic 11(1):3–17

Chapter 8
Strategizing FMV Conservation for Sustainable Agroecosystems in Rural-Urban Interface

8.1 Why to Retain FMVs and Traditional Agroecosystems

With approximately 9% of the world population and 14% of India's population, it has become a far-reaching goal to achieve "zero hunger" (SDG-2) of United Nations by 2030. Formulation of policy and implementation to provide food and nutritional security have always been having concerns of availability, accessibility and adequacy of food. The agricultural production systems have a wide range of internal and external factors affecting their structure and functions. Some of the factors include land scarcity, impacts of climate change, land degradation, socioeconomic conditions of farmers and unsustainable agricultural practices. For developing countries like India, strategic and localized approaches based on individual socioecological settings are required to move towards achieving food and nutritional security. Some practices of agricultural intensification have risk of weakening the ecological services provided by agrobiodiversity by affecting their abundance in and around fields (Diehl et al. 2012). Mechanization of agriculture, intensive monoculture and use of agrochemicals have jeopardized the agroecological integrity threatening the biodiversity rich traditional agricultural practices (Sardaro et al. 2016, Altieri et al. 2015; Singh and Singh 2017). Along with moving back to sustainable farming, ecological intensification such as wildlife-friendly farm management and biodiversity enhancement can actually contribute towards sustainable agricultural production (Pywell et al. 2015). In India, sustainable agriculture has been prioritized through policy initiatives like National Mission for Sustainable Agriculture (NMSA) which is one of the eight missions under National Action Plan for Climate Change. The mission defines sustainable agriculture as successful management of resources for agriculture, to satisfy the changing human need while maintaining ecological balance by avoiding depletion of natural resources. Traditional farmers are known as guardians of natural resources including agrobiodiversity and they have a better ability to overcome or cope with environmental and climatic risk (Chhatre and Agrawal 2008). Agroecosystems with rich biodiversity and significant traditional knowledge on indigenous practices in India have the potential to make the farmers less vulnerable (Maikhuri et al. 1996a, b; Maikhuri et al. 2001; Altieri 2002; Kuniyal et al. 2004; Vicziany and Plahe 2017). Traditional production systems are considered as robust production systems and have the capacity to withstand climate adversities to provide food and nutritional security in India (Patel et al. 2020; Watson 2019). There have been evidences of tree-based, traditional agricultural production system enhancing crop productivity and farm income by enhancing soil fertility, water management, pest and predator control, nutrient recycling and supporting the huge livestock population in India (Yadav et al. 2020; Kashyap et al. 2014; Murthy et al. 2013; Garrity et al. 2010; Kumar et al. 1998).

Agricultural production on fields in the wider area of megacities plays an important role for the supply of urban people with cereals, forages, and vegetables, but rapid urban growth constantly changes the conditions for crop production and exerts increasing pressure on farmers to adopt the cropping systems to the altered situation. Better knowledge about the nature, dynamics, and direction of changes in crop production is expected to support the decision making of stakeholders in planning processes of megacities and avoid negative repercussions of inappropriate decisions. Significant and rapid growth in urbanization in and around Bengaluru has brought about substantial changes in agriculture, natural vegetation including forest and field margins and water resources. Especially the productive agriculture lands of the rural areas of Bengaluru have been undergoing transformations which affect the quality, diversity, and scale of agricultural crop production and overall structure and composition of agroecosystems in rural-urban interface.

The result based on research, if compared chronologically, shows that overall vegetation cover has increased. The field visit or ground truth analysis supports the outcome of the work. Greening is mostly proliferated by government schemes to reclaim barren/waste land and agroforestry initiatives. Conservation of FMV and biological control practices is important because the recent research and studies held globally concludes FMV have been lost from the landscape and there is an annual net loss of field margin vegetation on earth. Conservation of FMV can aim at manipulating vegetation composition, structure and function to provide food resources and shelter from ecological disturbance. There is no published data on the abundance and diversity of vegetation along the field margins in the rural-urban transition region especially in the context of India. Inadequate studies and inconsistent result are responsible for not attracting attention into this arena.

The decrease in area under FMV in zones with expanding agricultural production (rural and transition zone) have been noticed and it could be due to introduction cash crops which has high demand in the city like Bengaluru (Table 8.1). Our field investigation and interactions with farmers revealed that the cash crops do not provide better production under shade hence the many native tree species which farmers used to raise in the field margins are no longer required. Another factor which was found responsible for decline in area of FMV is expansion of main land use for increasing the profit from cultivation of cash crops. The cash crop driven land use change has negatively affected the structure and functional characteristics of field margins and found all across the regions of northern transect of Bengaluru. The

Table 8.1 Understanding on field margin vegetation in rural-urban interface of Bengaluru, India

FMV dimensions	Major findings
FMV structure and area	Loss of FMV area (tree cover) from 2005 to 2017. Lower loss of FMV area in urban zone (4.6%) as compared to 12–16% in other 2 zones
FMV composition	Rich transition zone with 76 species
Importance of FMV	Economic value under 12 categories, low knowledge on ecological importance. Monetary benefits only from perennial edible producing trees are
TEK on FMV	10–60% HHs have TEK on 21, 18 and 12 species of trees, shrubs and herbs respectively. Traditional use of FMV/FMV products in 20 different medicinal use categories
LULCC	Increase in built-up area, plantation and open forest; whereas decrease in agricultural land, barren land and waterbodies
Vegetation	Both the vegetation indices (NDVI and SAVI) show significant increase in vegetation cover
Crop-FMV relationship	Cereals crops (Ragi and maize) boundaries are species rich with 13 tree species. Clearing of FMV tree in grapes, lawn grass and flower crop plots
FMV area mapping and monitoring	The methods validated and discussed in Chap. 6 provides an accuracy (87%) of acceptable level to further use, upscale and improvise for mapping and monitoring of field margin vegetations

result based on research, if compared chronologically, shows that overall vegetation cover has increased. The field visit or ground truth analysis supports the spatio-temporal outcomes in this endeavour. Conservation of FMV and biological control practices are important because the recent research and studies held globally concludes that the areas under FMV have been lost from the landscape and there is an annual net loss of ecological services from the field margin vegetation on the earth. Conservation of FMV can aim at manipulating vegetation composition, structure and function to provide food resources and shelter from ecological disturbance. There is no published data on the abundance and diversity of vegetation along the field margins in the rural-urban transition region especially in the context of India. Inadequate studies and inconsistent result are responsible for not attracting attention into this arena.

Mapping and monitoring provide significant inputs for environmental planning. Upscaling of results of mapping and delineation FMVs should be done extensively. Based on the results and their accuracy, it is suggestive that the method is scalable for identifying, assessing and managing FMVs for sustainable socioecological development. The additive effects of all the components into the field have to be investigated and special attention and care need to be promised by the stakeholders. The combination of local ecological and cultural knowledge of people and scientific landscape

Photo 8.1 Mixed traditional horticulture plot with diversity of FMV in northern transect of Bengaluru

management is the most promising strategy for strengthening the ecosystem services from the landscape dominated by agriculture. Regular monitoring and maintenance of field margin networks and its potential usage can result high quality of landscape. Interventions in vegetative components of naturally regenerated filed margin can increase its potential for agricultural landscape sustainability. In India, role of FMV is not properly understood and therefore, there is a need for in-depth research in this aspect so that appropriate solutions can be provided for sustainable socio-ecological development of the landscape dominated by agriculture (Photo 8.1)

8.2 Strategies for Developing Sustainable Agroecosystems Through Conserving FMV

The discussion in previous chapters of this book provide a preliminary understanding on the changing field margin vegetation with respect to changing agroecosystems in rural-urban interface. It has also brought in an overview of economic and ecological importance of FMVs, their use and knowledge possessed by the farmers which is based on authentic empirical data and information. It is seen that traditional structure of boundary vegetation has been replaced by economically important exotic species, which is more prevalent in rural and transitional sections of the transect where policy induced agroforestry practices are more extensive. Farmers are well aware of the importance of traditional species in their farmland; but the care should be taken to replace those species for achieving economic goals which may affect the ecosystem balance as well as sustainable income in the long run. Policies should be formulated to encourage agricultural production as a part of the ecosystem which will enable to conserve the non-crop biodiversity at the field margin and their associated knowledge. In this approach, identification and valuing of ecosystem services of all the components of agroecosystem should be carried out. Development of policy should be by all the departments (agriculture, forest, water resource, urban and rural development) in synchronization to deal with adverse impacts of landscape change. To contribute towards robust policy for preserving agrobiodiversity and traditional knowledge in highly transitional landscape like rural-urban interface should prioritize the following agenda as future work.

1. Systematic identification of the species in field margins and estimation of ecosystem services of all species in field boundary.
2. Detailed investigation and identification of keystone species for specific agroecosystem and review of agroforestry policy based on the findings.
3. Economic and ecological valuation of field margin vegetations.
4. Recognition of agroccosystem which has explicit inclusion of non-crop diversity.
5. Identification of factors affecting the agroecosystem along the rural-urban gradient to develop strategies for their control so that sustainable agricultural practices can be encouraged.
6. Awareness programmes with farmers on environmental importance of field margin vegetation with important species. Encouragement for developing semi-natural field boundaries.
7. Interventions in vegetative components of naturally regenerated filed margins can increase its potential for agricultural landscape sustainability.
8. Multidisciplinary and participatory research on traditional ecological knowledge to establish them in mainstream research and contemporary discussions (Photo 8.2).

Emphasis of socio-cultural set-up and documentation of traditional knowledge on the agrobiodiversity will be helpful in conservation of traditional agrobiodiversity (Nautiyal et al. 2008) which will subsequently help the communities to cope with the adverse impacts of environmental changes including climate change. Rai (2007) also identifies the significant role of traditional knowledge and institutions in sustainable resource use and management. Indigenous agricultural and resource management practices are obtained from parents, nature, friends, neighbours which are crucial for sustaining location specific ecology and farming system (Singh and Sureja 2006). This knowledge gets enhanced till the traditional practices are in place, but starts eroding once the agroecosystems undergo transformation due to drivers such as urbanization, market demand etc. Aichi Biodiversity Target 18 and India's National

Photo 8.2 Agroforestry (Guava-Silver oak) in the transition zone of northern Bengaluru

Biodiversity Target under Convention on Biodiversity have recognized the importance of traditional knowledge, its protection and development for economic benefits (Jasmine et al. 2016). Propagation and scientific establishment of the knowledge to conserve the traditional agroecosystems of the study area through strategic policy framework have potential to contribute towards sustainable socioecological development. TEK systems are acknowledged as potential contributor in decision making for adaptation to climate change (Turner and Spalding 2013) and are identified as important for maintaining the field margin vegetation as part of agroecosystem, a combined approach for conservation of TEK in FMV is essential for sustainable socioecological development.

8.3 Way Forward

Three has been changes in structure, function and composition of field margin vegetations across the transect. The most pertinent questions arising as a conclusion to all the findings and proceeding discussions are that which is the next step, what is the most important concern to address, how to deal the inevitable changes and how to advance towards achieving all those. The results and understanding suggest further detailed study the role and economic value of FMV in the socioecological system, drivers of change to formulate management strategy. The major gaps in research are identified as:

- Ecological and economic value of FMV to be quantified and understood comprehensively. Methodology needs to be developed for non-economic attributes.
- Drivers of change are not identified with the extent of impact.
- Pattern and impact of urbanization in different peripheral sides of the city is not similar.
- Upscaling of RS-GIS based effective methodology to identify and delineate FMV which will less tedious and time consuming.
- Push–Pull paradigm to define the transitional landscape and its impact to be assessed.
- Need comprehensive assessment of FMV as a significant part of agroforestry system in the context of climate change.

References

Altieri MA (2002) Agroecology: the science of natural resource management for poor farmers in marginal environments. Agric Ecosyst & Environ 93(1–3):1–24

Altieri MA, Nicholls CI, Henao A, Lana MA (2015) Agroecology and the design of climate change-resilient farming systems. Agron Sustain Dev 35(3):869–890. https://doi.org/10.1007/s13593-015-0285-2

Chhatre A, Agrawal A (2008) Forest commons and local enforcement. Proc Natl Acad Sci 105(36):13286–91

Diehl E, Wolters V, Birkhofer K (2012) Arable weeds in organically managed wheat fields foster carabid beetles by resource-and structure-mediated effects. Arthropod-Plant Inter 6(1):75–82. https://doi.org/10.1007/s11829-011-9153-4

Garrity DP, Akinnifesi FK, Ajayi OC, Weldesemayat SG, Mowo JG, Kalinganire A, Larwanou M, Bayala J (2010) Evergreen agriculture: a robust approach to sustainable food security in Africa. Food Secur 2(3):197–214

Jasmine B, Singh Y, Onial M, Mathur VB (2016) Traditional knowledge systems in India for biodiversity conservation. Indian J Tradit Knowl 15(2)

Kashyap SD, Dagar JC, Pant KS, Yewale AG (2014) Soil conservation and ecosystem stability: natural resource management through agroforestry in Northwestern Himalayan Region. In: Dagar J, Singh A, Arunachalam A (eds) Agroforestry systems in India: Livelihood Security & Ecosystem Services. Advances in agroforestry, vol 10. Springer, New Delhi. https://doi.org/10.1007/978-81-322-1662-9_2

Kumar BM, George SJ, Jamaludheen V, Suresh TK (1998) Comparison of biomass production, tree allometry and nutrient use efficiency of multipurpose trees grown in woodlot and silvopastoral experiments in Kerala India. For Ecol Manag 112(1–2):145–163

Kuniyal JC, Vishvakarma SC, Singh GS (2004) Changing crop biodiversity and resource use efficiency of traditional versus introduced crops in the cold desert of the Northwestern Indian Himalaya: a case of the Lahaul valley. Biodivers Conserv 13(7):1271–1304

Maikhuri RK, Rao KS, Saxena KG (1996) Traditional crop diversity for sustainable development of Central Himalayan agroecosystems. Int J Sustain Dev World Ecol 3(3):8–31. https://doi.org/10.1080/13504509609469926

Maikhuri RK, Rao KS, Saxena KG (1996) Traditional crop diversity for sustainable development of Central Himalayan agroecosystems. Int J Sustain Dev World Ecol 2:1–24

Maikhuri RK, Rao KS, Semwal RL (2001) Changing scenario of Himalayan agroecosystems: loss of agrobiodiversity, an indicator of environmental change in Central Himalaya, India. Environmentalist 21(1):23–39. https://doi.org/10.1023/A:1010638104135

Murthy IK, Gupta M, Tomar S, Munsi M, Tiwari R, Hegde GT, Ravindranath NH (2013) Carbon sequestration potential of agroforestry systems in India. J Earth Sci Clim Change 4(1):1–7. https://doi.org/10.4172/2157-7617.1000131

Nautiyal S, Bisht V, Rao KS, Maikhuri RK (2008) The role of cultural values in agrobiodiversity conservation: a case study from Uttarakhand, Himalaya, J Hum Ecol 23(1):16. https://doi.org/10.1080/09709274.2008.11906047

Patel SK, Sharma A, Singh GS (2020) Traditional agricultural practices in India: an approach for environmental sustainability and food security. Energy Ecol Environ 1–19. doi:https://doi.org/10.1007/s40974-020-00158-2

Pywell RF, Heard MS, Woodcock BA, Hinsley S, Ridding L, Nowakowski M, Bullock JM (2015) Wildlife-friendly farming increases crop yield: evidence for ecological intensification. Proc R Soc B: Biol Sci 282(1816):20151740. https://doi.org/10.1098/rspb.2015.1740

Rai SC (2007) Traditional ecological knowledge and community-based natural resource management in northeast India. J Mt Sci 4(3):248–258

Sardaro R, Girone S, Acciani C, Bozzo F, Petrontino A, Fucilli V (2016) Agro-biodiversity of Mediterranean crops: farmers' preferences in support of a conservation programme for olive landraces. Biol Conserv 201:210–219. https://doi.org/10.1016/j.biocon.2016.06.033

Singh R, Singh GS (2017) Traditional agriculture: a climate-smart approach for sustainable food production. Energy Ecol Environ 2(5):296–316. https://doi.org/10.1007/s40974-017-0074-7

Singh RK, Sureja AK (2006) Community knowledge and sustainable natural resources management: learning from the Monpa of Arunachal Pradesh. TD: J Trans Res South Afr 2(1):73–102

Turner N, Spalding P (2013) "We might go back to this"; drawing on the past to meet the future in Northwestern North American Indigenous communities. Ecol Soc 18(4)

Vicziany M, Plahe J (2017) Food security and traditional knowledge in India: the issues. South Asia 40(3):566–581. https://doi.org/10.1080/00856401.2017.1342181

Watson D (2019) Adaptation to climate change through adaptive crop management. In: Sarkar A, Sensarma S, vanLoon G (eds) Sustainable solutions for food security. Springer, Berlin. 191–210. https://doi.org/10.1007/978-3-319-77878-5_10

Yadav OP, Singh JP, Kakani RK, Mahla HR, Rajora MP, Singh A, Meghwal PR, Verma A (2020) Managing agrobiodiversity of Indian drylands for climate-change adaptation. Indian J Plant Genetic Resour 33(1):3–16

Annexure A: Tree Species in Six Villages of the Northern Transect of Bengaluru

See Table A.1.

Table A.1 Tree species listed in six villages of the northern transect of Bengaluru

S. No.	Tree species			Rural village		Transition (PeriUrban) Village		Urban village	
	Common name	Scientific name	Family	Heggadahalli	Muddenahalli	Kundana	Konaghatta	Singanayakanahalli	Arekere
1	Cocount tree	Cocos nucifera L.	Arecaceae	+	+	+	+	+	+
2	Indian beech tree	Pongamia pinnata (L.) Pierre	Fabaceae	+	+	+	+	+	-
3	Tamarind	Tamarindus indica L.	Fabaceae	+	+	+	+	+	+
4	Guava	Psidium guajava L.	Myrtaceae	+	+	+	+	+	+
5	Teak	Tectona grandis L.	Verbenaceae	+	+	+	+	+	+
6	Jakfruit	Artocarpus heterophyllus Lam	Moraceae	+	+	+	+	+	+
7	Neem tree	Azadirachta indica A. Juss	Meliaceae	+	+	+	-	+	+
8	Eucalyptus tree	Eucalyptus spp.	Myrtaceae	+	+	+	+	+	+
9	Silver Oak	Grevillea robusta	Proteaceae	+	+	+	+	-	+
10	Mango tree	Mangifera indica L.	Anacardiaceae	+	+	+	+	+	+
11	Acacia tree	Acacia spp. L.	Fabaceae	+	-	-	+	-	-

(continued)

Table A.1 (continued)

S. No.	Tree species			Rural village		Transition (PeriUrban) Village		Urban village	
	Common name	Scientific name	Family	Heggadahalli	Muddenahalli	Kundana	Konaghatta	Singanayakanahalli	Arekere
12	Cluster fig tree	*Ficus racemosa* L.	Moraceae	-	-	-	+	-	-
13	Jamun	*Syzygium cumini* (L.) Skeels	Myrtaceae	-	-	+	+	-	+
14	Banyan tree	*Ficus benghalensis* L.	Moraceae	-	-	+	+	-	-
15	Cashew tree	*Anacardium occidentale* L.	Anacardiaceae	-	-	+	+	-	-
16	Lemon tree	*Citrus limon* (L.) Osbeck	Rutaceae	-	-	-	+	-	+
17	Castor tree	*Ricinus communis* L.	Euphorbiaceae	-	-	-	+	+	-
18	Gulmohur tree	*Delonix regia* (Boj. ex Hook.) Raf.	Fabaceae	-	-	+	+	-	-

(continued)

Table A.1 (continued)

S. No.	Tree species			Rural village		Transition (PeriUrban) Village		Urban village	
	Common name	Scientific name	Family	Heggadahalli	Muddenahalli	Kundana	Konaghatta	Singanayakanahalli	Arekere
19	Sapota	*Manilkara zapota* (L.) P. Royen	Sapotaceae	-	-	+	+	+	-
20	Curry plant	*Murraya koenigii* (L.) Sprengel	Rutaceae	-	-	-	+	+	-
21	Jamaica Cherry	*Muntingia calabura* L.	Muntingiaceae	-	-	+	-	-	-
22	Custard apple	*Annona squamosa* L.	Annonaceae	-	-	+	-	-	-
23	Peepal tree,	*Ficus religiosa* L.	Moraceae	-	-	+	+	-	+
24	Rain tree	*Samanea saman* (Jacq.) Merr	Fabaceae	-	-	+	-	-	-
25	Wild date palm	*Phoenix sylvestris* (L.) Roxb	Arecaceae	-	-	+	-	-	-
26	Papaya	*Carica papaya* L.	Caricaceae	-	-	+	-	-	-

(continued)

Table A.1 (continued)

S. No.	Tree species			Rural village		Transition (PeriUrban) Village		Urban village	
	Common name	Scientific name	Family	Heggadahalli	Muddenahalli	Kundana	Konaghatta	Singanayakanahalli	Arekere
27	Popcorn Bush Cedar	*Cassia spectabilis* DC	Fabaceae	-	-	+	-	-	-
28	Pomegranate	*Punica granatum* L.	Punicaceae	-	-	+	-	-	+
29	Indian gooseberry	*Phyllanthus emblica* L.	Phyllanthaceae	-	-	-	-	-	+
Total				11	10	23	20	12	14

Annexure B: Shrub Species in Six Villages of the Northern Transect of Bengaluru

See Table B.1.

Table B.1 Shrub species listed in six villages of the northern transect of Bengaluru

S.No.	Tree Species			Rural village		Peri_urban village		Urban village	
	Common name	Scientific name	Family	Heggadahalli	Muddenahalli	Kundana	Konaghatta	Singanayakanahalli	Arekere
1	Gigantic swallow wort	*Calotropis gigantea* (L.) *R.Br*	Apocynaceae	+	+	+	+	+	+
2	Wild-sage	*Lantana camara* L.	Verbenaceae	+	+	+	+	+	+
3	Congress weed	*Parthenium hysterophorus* L.	Asteraceae	+	+	+	+	+	+
4	Country mallow	*Sida cordifolia* L.	Malvaceae	+	+	+	+	+	+
5	Queensland-hemp	*Sida rhombifolia* L.	Malvaceae	+	-	+	+	-	+
6	Siam weed	*Chromolaena odorata* L.	Asteraceae	+	+	+	+	-	-
7	Clotbur	*Xanthium indicum* L.	Asteraceae	+	-	+	+	-	-
8	wire weed	*Sida acuta* L.	Malvaceae	+	+	+	+	+	+
9	Wild eggplant	*Solanum torvum* Sw.	Solanaceae	+	+	+	+	+	+
10	Diamond burbark	*Triumfetta rhomboidea* Jacq	Malvaceae	+	-	-	-	-	+
11	Indian mallow	*Abutilon indicum* (L.) sweet	Malvaceae	-	+	+	+	-	-
12	Castor-oil-plant	*Ricinus communis* L.	Euphorbiaceae	-	+	+	+	-	-
13	Asian pigeonwings	*Clitoria ternatea* L.	Fabaceae	-	-	-	+	-	-
14	Golden trumpet	*Allamanda cathartica* L.	Apocynaceae	-	-	+	-	-	-
15	Yellow oleander	*Thevetia peruviana* L.	Apocynaceae	-	-	+	-	-	-
16	Barbados nut	*Jatropha curcas* L.	Euphorbiaceae	-	-	+	-	-	-
17	Great bougainvillea	*Bougainvillea spectabilis* Willd	Nyctaginaceae	-	-	+	-	-	-

(continued)

Table B.1 (continued)

| S.No. | Tree Species | | | Rural village | | Peri_urban village | | Urban village | |
	Common name	Scientific name	Family	Heggadahalli	Muddenahalli	Kundana	Konaghatta	Singanayakanahalli	Arekere
18	Erect prickly pear	*Opuntia dilleni* (Ker-Gawl.) Haw	Cactaceae	-	-	+	-	-	-
19	Caesarweed	*Urena lobata* L.	Malvaceae	-	-	-	-	+	-
Total				10	9	15	12	7	8

Annexure C: Herb Species Listed in Six Villages of the Northern Transect of Bengaluru

See Table C.1.

S. Nautiyal et al., *Field Margin Vegetation and Socio-Ecological Environment*,
Environmental Science and Engineering,
https://doi.org/10.1007/978-3-030-69201-8

Table C.1 Herb species listed in six villages of the northern transect of Bengaluru

| S. No | Tree Species | | | Rural Village | | Peri_Urban Village | | | Urban Village | |
	Common name	Scientific name	Family	Heggadahalli (R1)	Muddenahalli (R2)	Kundana (P1)	Konaghatta (P2)	Singanayakanahalli (U1)	Arekere (U2)
1	Chinese Wedelia	Wedelia chinensis L.	Asteraceae	+	+	+	+	+	+
2	Touch-me-not	Mimosa pudica L.	Fabaceae	+	+	+	+	+	+
3	Nodeweed	Synedrella vialis L.	Asteraceae	+	+	+	+	-	-
4	Coatbuttons	Tridax procumbens L.	Asteraceae	+	+	+	+	+	+
5	Hygrophila	Asteracantha longifolia L.	Acanthaceae	+	+	+	+	-	-
6	Green amaranth	Amaranthus viridis L.	Amaranthaceae	+	+	+	+	+	+
7	Sleeping beauty	Oxalis corniculata L.	Oxalidaceae	+	+	+	+	-	-
8	Hairy fleabane	Erigeron bonariensis L.	Asteraceae	+	+	+	+	-	-
9	Shrubby false Button	Spermacoce verticillata	Rubiaceae	+	+	-	+	+	-
10	False buttonweed	Spermacoce articularis L.	Rubiaceae	+	+	+	+	-	-
11	Rattleweed	Crotalaria retusa L.	Fabaceae	+	+	+	+	+	+

(continued)

Table C.1 (continued)

S. No	Tree Species			Rural Village		Peri_Urban Village		Urban Village	
	Common name	Scientific name	Family	Heggadahalli	Muddenahalli	Kundana	Konaghatta	Singanayakanahalli	Arekere
				(R1)	(R2)	(P1)	(P2)	(U1)	(U2)
12	Joyweed	Alternanthera sessilis L.	Amaranthaceae	+	+	+	+	+	+
13	Batchelor's button	Gomphrena celosioides	Amaranthaceae	+	+	+	+	+	+
14	Thumbai	Leucas aspera (Wild.) Spreng	Lamiaceae	+	+	+	+	-	+
15	Asthma-plant	Euphorbia hirta L.	Euphorbiaceae	+	+	+	+	+	+
16	Fire plant	Euphorbia heterophylla L.	Euphorbiaceae	+	+	-		-	+
17	White-eye	Richardia brasiliensi Gomes.	Rubiaceae	+	+	-	+	-	+
18	Spider weed	Cleome spp.	Cleomaceae	+	-	-	-	-	-
19	Horse weed	Conyza canadensis L.	Asteraceae	+	+	-	+	-	-
20	Stonebreaker	Phyllanthus amarus L.	Phyllanthaceae	-	+	+	+	-	+

(continued)

Table C.1 (continued)

S. No	Tree Species			Rural Village		Peri_Urban Village		Urban Village	
	Common name	Scientific name	Family	Heggadahalli	Muddenahalli	Kundana	Konaghatta	Singanayakanahalli	Arekere
				(R1)	(R2)	(P1)	(P2)	(U1)	(U2)
21	Ban Tulasi	*Croton bonplandianum* Baill	Euphorbiaceae	-	+	+	+	+	-
22	Little Ironweed	*Cyanthilium cinereum* H.Rob	Asteraceae	-	+	-	+	-	-
23	Girdlepod	*Mitracarpus hirtus* (L.) DC	Rubiaceae	-	+	-	+	+	+
24	White weed	*Ageratum conyzoides* L.	Asteraceae	-	+	-	-	+	+
25	Climbing Day flower	*Commelina diffusa* Burm.f	Commelinaceae	-	+	-	-	+	-
26	Rosy periwinkle	*Catharanthus roseus* (L.) G. Don	Apocynaceae	-	-	+	-	+	-
27	Elephant-ear	*Colocasia* spp.	Araceae	-	-	+	-	-	-
28	Marigold	*Tagetes erecta* L.	Asteraceae	-	-	+	-	-	-

(continued)

Table C.1 (continued)

| S. No | Tree Species | | | Rural Village | | Peri_Urban Village | | Urban Village | |
	Common name	Scientific name	Family	Heggadahalli (R1)	Muddenahalli (R2)	Kundana (P1)	Konaghatta (P2)	Singanayakanahalli (U1)	Arekere (U2)
29	Tomato	*Solanum lycopersicum* L.	Solanaceae	-	-	+	-	-	-
30	Banana	*Musa* spp.	Musaceae	-	-	+	-	-	+
31	Zinnia	*Zinnia* spp.	Asteraceae	-	-	+	-	-	-
Total				19	24	22	21	14	15

Annexure D: Percentage of Farms Having Tree in the Field Margins for Different Crop Groups and Presence of FMV Tree Species (From the Selected Fields of the Northern Transect of Bengaluru)

See Table D.1.

Table D.1 Percentage of farms having tree in the field margins for different crop groups and presence of FMV tree species in the northern transect of Bengaluru

Crop class	Rural			Transition			Urban		
	Crops	%of plots with FMV	FMV-Tree species	Crops	%of plots with FMV	FMV-Tree species	Crops	%of plots with FMV	FMV-Tree species
Cereals and pulses	Ragi, Maize Togari	41.67	*Mangifera indica, Ziziphus jujuba, Annona squamosa, Artocarpus heterophyllus, Cocos nucifera, Ficus benghalensis*	Finger millet, Redgram, Sorghum, Horsegram, and Mustard	61.90	*Grevillea robusta, Eucalyptus* spp., *Tectona grandis, Cocos nucifera, Pongamia pinnata, Ficus benghalensis, Phoenix dactylifera*	Ragi, maize, lablab, horsegram	57.14	*Ficus benghalensis, Pongamia pinnata, Bauhinia variegata, Melia dubia*
Vegetables	Snake gourd, tomato, brinjal, cabbage, cauliflower	60	*Artocarpus heterophyllus, Pongamia pinnata, Cocos nucifera, Eucalyptus* spp.	Cucumber, corriander, chilli, tomato, ginger, beans,	27.27	*Cocos nucifera, Ficus benghalensis*	Beans, bitter gourd, cauliflower, tomato	57.14	*Tectona grandis, Pongamia pinnata, Cocos nucifera*
Plantation	Arecanut, mango	100	*Tectona grandis, Grevillea robusta, Cocos nucifera*	Sapote, coconut, teak	50.00	*Grevillea robusta*	Guava, Mango, Pomegranate	71.43	*Pongamia pinnata, Cocos nucifera, Eucalyptus* spp., *Artocarpus heterophyllus*

(continued)

Table D.1 (continued)

Crop class	Rural			Transition			Urban		
	Crops	%of plots with FMV	FMV-Tree species	Crops	%of plots with FMV	FMV-Tree species	Crops	%of plots with FMV	FMV-Tree species
Grapes		41.67	*Cocos nucifera*		0			50%	*Tectona grandis, Grevillea robusta, Mangifera indica, Azadirchta indica, Cocos nucifera*
Flower	Rose, marigold, lilly	60%	*Mangifera indica, Ficus benghalensis*	Rose, marigold	66.67	*Azadirachta indica, Grevillea robusta*	Rose	0	
Lawn grass					0			0	

Annexure E: Relative Density of Herb Species in Six Study Villages of the Northern Transect of Bengaluru

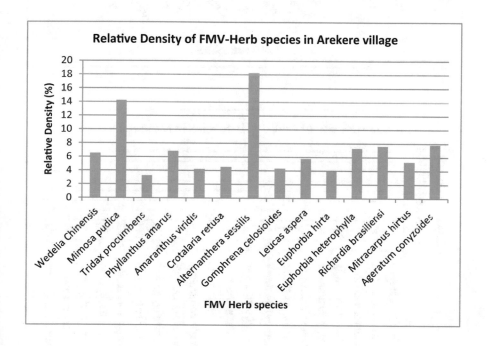

S. Nautiyal et al., *Field Margin Vegetation and Socio-Ecological Environment*,
Environmental Science and Engineering,
https://doi.org/10.1007/978-3-030-69201-8

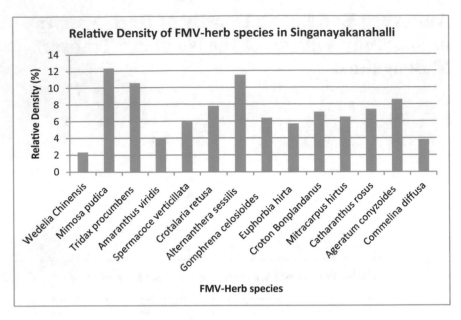

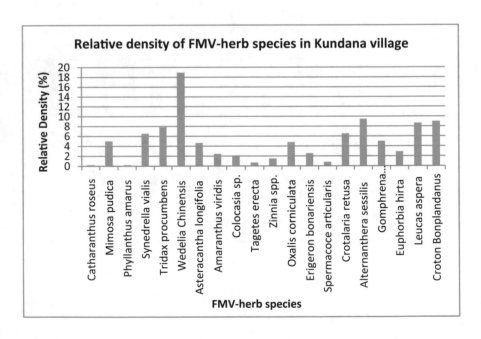

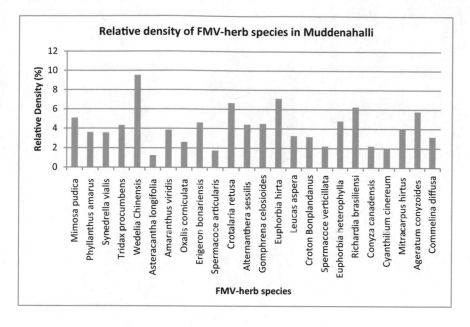

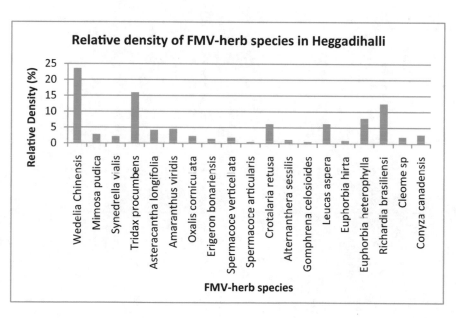

Annexure F: Relative Frequency of FMV-herb Species in Six Study Villages of the Northern Transect of Bengaluru

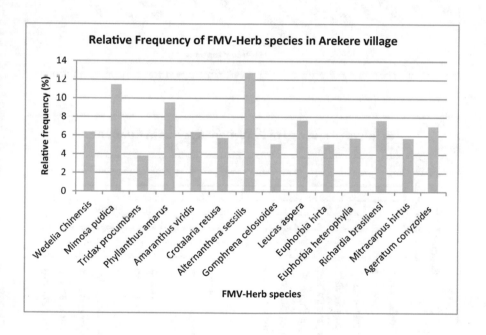

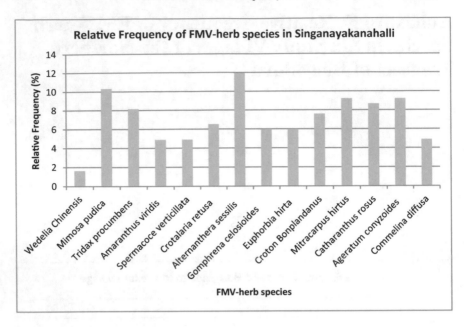

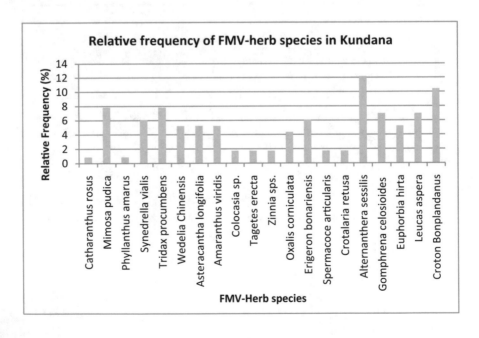

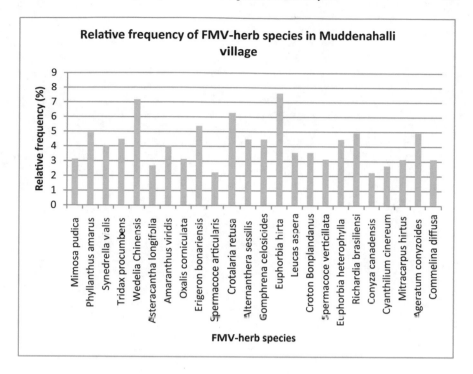

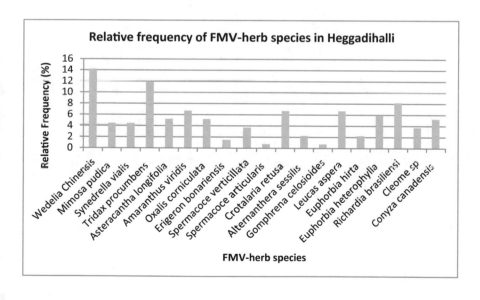

Annexure G: Photo Plates-Some FMV Species Prevalent in the Northern Transect of Bengaluru

Tagetes erecta

Basella alba

Ficus benghalensis

Blumea spp.

Amaranthus blitum

Ocimum sanctum

Ricinus communis   Mangifera indica   Agave americana

Napier grass   Eucalyptus glabrus   Xanthium indicum

Euphorbia heterophylla   Richardia scabra   Commelina benghalensis

Hyptis suaveolens   Punica granatum   Capsicum annuum

Leucas aspera   *Boerhavia diffusa*   *Beta vulgaris*

Ficus spp.   *Tectona grandis*   *Calotropis gaigantia*

Tridax procumbens   *Brassica nigra*   *Mimosa pudica*

Bambusa vulgaris   *Tephrosia purpurea*   *Oxalis corniculata*

Rhynchelytrum repens   *Jasminum sambac*   *Solanum nigrun*

Lantana camara   *Kyllinga nemoralis*   *Manilkara zapota*

Blumea lacera   *Syzygium cumuni*   *Ipomoea staphylina*

Croton bonplandianum   *Ipomoea obscura*   *Eupatorium* spp.

Indigofera cardifolia   *Synedrella nodiflora*   *Cardiospermum halicacabum*

Cassia spectabilis   *Gravillea robusta*   *Morus alba*

Cassia tora   *Alternanthera pungens*   *Phoenix sylvestris*

Jatropha curcas

Annexure H: Correlation Between FMV Products and their Economic Benefits in the Northern Transect of Bengaluru

© The Editor(s) (if applicable) and The Author(s), under exclusive license
to Springer Nature Switzerland AG 2021
S. Nautiyal et al., *Field Margin Vegetation and Socio-Ecological Environment*,
Environmental Science and Engineering,
https://doi.org/10.1007/978-3-030-69201-8

Vairables	Agricultural land-past	Agricultural Land-	No. of crops- past	No. of crops-present	Annual income	No. of livestock-past	No. of livestock-present	No. of useful FMV	Wood from forest-past	Fodder from forest-past	Leaf litter from forest-	Other resources from	Wood from forest-	Fodder from forest-	Leaf litter from forest-	Other resources from	Leaf litter from FMV-	Wood from FMV-past	Fodder from FMV-past	other resources from	Leaf litter from FMV-	Wood from FMV-	Fodder from FMV-	Other resources from
Agricultural land-past	1																							
Agricultural Land-present	0.36	1.00																						
No. of crops- past	0.48	0.10	1.00																					
No. of crops-present	0.16	0.60	0.15	1.00																				
Annual income	0.30	0.05	0.26	0.24	1.00																			
No. of livestock-past	0.31	0.05	0.06	0.01	0.10	1.00																		
No. of livestock-present	0.32	0.07	0.05	0.06	0.16	0.38	1.00																	
No. of useful FMV species	0.003	-0.06	0.00	0.08	0.24	0.07	-0.10	1.00																
Wood from forest-past	-0.01	0.13	0.36	0.05	-0.10	0.16	0.03	-0.08	1.00															
Fodder from from forest-past	-0.01	-0.04	0.06	0.13	-0.17	0.11	0.30	-0.08	0.41	1.00														
Leaf litter from forest-past	0.11	0.16	0.29	0.05	-0.19	0.07	0.03	0.01	0.46	0.33	1.00													
Other resources from forest-past	0.02	0.00	-0.34	-0.06	0.33	0.05	-0.00	0.07	0.45	0.38	0.87	1.00												
Wood from forest-present	-0.10	-0.10	0.08	-0.16	-0.01	-0.05	0.06	-0.03	0.36	0.03	0.05	0.07	1.00											
Fodder from forest-present	0.03	0.07	-0.03	-0.13	0.08	0.13	0.54	-0.07	0.09	0.37	0.21	0.24	0.38	1.00										

Leaf litter from forest-present	-0.09	-0.10	-0.12	-0.11	-0.10	-0.06	-0.10	-0.15	0.07	-0.03	-0.05	-0.06	0.10	0.04	1.00								
Other resources from forest-present	-0.01	0.04	0.04	-0.02	0.12	0.04	-0.13	0.05	0.06	0.11	0.24	0.26	0.05	0.45	0.02	1.00							
Leaf litter from FMV-past	-0.09	0.05	0.25	-0.01	0.20	0.05	0.09	0.01	0.38	0.38	0.87	0.85	-0.11	0.09	-0.05	-0.06	1.00						
Wood from FMV-past	0.15	0.11	0.45	-0.01	0.30	0.06	0.17	0.12	0.46	0.30	0.64	0.68	-0.05	0.26	0.07	0.07	0.71	1.00					
Fodder from FMV-past	0.01	-0.11	0.11	0.16	-0.21	0.03	0.09	0.14	0.27	0.52	0.59	0.62	-0.10	0.05	-0.06	0.14	0.66	0.52	1.00				
other resources from FMV-past	-0.07	-0.08	0.36	-0.11	0.30	-0.07	-0.14	0.05	0.44	0.25	0.57	0.31	-0.18	-0.04	-0.05	0.70	0.61	0.60	1.00				
Leaf litter from FMV-present	-0.05	-0.02	-0.07	-0.03	0.38	0.08	0.05	0.30	-0.19	-0.17	0.10	0.13	-0.29	0.13	0.04	0.05	0.11	0.14	0.11	0.09	1		
Wood from FMV-present	0.02	0.03	0.23	-0.04	0.27	-0.08	0.04	0.13	0.15	0.15	0.41	0.32	0.07	0.13	-0.06	0.42	0.23	0.25	0.34	0.18	0.09	1	
Fodder from FMV-present	0.15	0.09	0.17	-0.10	0.14	0.00	0.19	0.27	-0.06	0.29	0.07	0.09	0.02	0.07	-0.09	0.06	0.08	0.03	0.26	0.05	0.02	0.22	1

Bibliography

Aavik T, Liira J (2010) Quantifying the effect of organic farming, field boundary type and landscape structure on the vegetation of field boundaries. Agric Ecosyst Environ 135(3):178–186

Aswani S, Lemahieu A, Sauer WHH (2018) Global trends of local ecological knowledge and future implications. PLoS ONE 13(4):e0195440

Aswani, S, Lemahieu A, Sauer W (2018) Global trends of local ecological knowledge and future implications, PLoS

Audrey A (2018) Two decades of change in a field margin vegetation metacommunity as a result of field margin structure and management practice changes. Agric Ecosyst Environ 251:1–10

Baselga A (2013) Separating the two components of abundance-based dissimilarity: balanced changes in abundance vs. abundance gradients. J Ecol Evol 4:552–557

Dormann CF, Schweiger O, Augenstein I, Bailey D, Billeter R, De Blust G, DeFilippi R, Frenzel M, Hendrickx F, Herzog F (2007) Effects of landscape structure and land-use intensity on similarity of plant and animal communities. J Global Ecol. Biogeogr 16:774–787

El-Gammal MI, Ali RR, Samra RA (2014) NDVI threshold classification for detecting vegetation cover in Damietta governorate. Egypt. Am Sci 10:8

Gómez-Baggethun E (2013 Dec 1) EsteveCorbera, Victoria Reyes-García (2013) Traditional ecological knowledge and global environmental change. research findings and policy implications. Ecol Soc 18(4):72

Gilbert B, Lechowicz MJ (2004) Neutrality, niches, and dispersal in a temperate forest understory. Proc Natl Acad Sci U S a 101:7651–7656

Gokhale Y, Negi AK eds (2011) Community-based biodiversity conservation in the Himalayas. The Energy and Resources Institute (TERI)

Gómez-Baggethun E, Reyes-García V (2013) Reinterpreting change in traditional ecological knowledge. Hum Ecol Interdiscip J 41(4):643–647

Gómez-Baggethun E, Reyes-García V (2013) Reinterpreting change in traditional ecological knowledge. Hum Ecol Interdiscip J 41(4). https://doi.org/10.1007/s10745-013-9577-9

Grimm NB, Faeth SH, Golubiewski NE, Redman CL, Wu J, Bai X, Briggs JM (2008) Global change and the ecology of cities. Sci 319(5864):756–760

Harisha RP, Padmavathy S, Nagaraja BC (2015) Traditional ecological knowledge (TEK) and its importance in south India: perspective from local communities. Traditional ecological knowledge (TEK) and its importance in South India 14(1):311–326

Harishkumar HV, Reddy BV (2017) Impact of urbanization on Land use pattern of rural-urban gradient of Bengaluru north: an economic analysis. Econ Aff 62(2):303–312

Holland JM, Luff ML (2000) The effects of agricultural practices on Carabidae in temperate agroecosystems. J Integr Pest Manag 5(2):109–129

Holland JM (2007) Managing uncropped land in order to enhance biodiversity benefits of the arable farmed landscape. Ann Appl Biol 81:255–260

Huete AR (1988) A soil-adjusted vegetation index (SAVI). Remote Sens Environ 25:295–309

Jagadeesh CB (2015) Dynamics of rapid urbanization of Bengaluru and its impact on landuse/land-cover—a case study of Vrishabhavathi sub-watershed. Int J Eng Technol 2(3)

Kumara BH, Babu KL (2009) Traditional knowledge system (medicine): a case study of Arakalgud Taluk, Karnataka. Institute for Social and Economic Change, India

Maikhuri RK, Nautiyal S, Rao KS, Saxena KG (1998) Role of medicinal plants in the traditional health care system: a case study from Nanda Devi Biosphere Reserve. Curr Sci 75:152–157

Marshall EJ (1987) Field margin flora and fauna: interaction with agriculture. In: Way JM, Greig-Smith PJ (eds) Monograph No. 35. Field Margins. British Crop Protection Council, Thornton Heath, Surrey, pp 23–33

Marshall EJ (1996) Factors affecting the floral diversity in European field margin networks. In: Simpson I, Dennis P (eds) The spatial dynamics of biodiversity, pp 97–104

Marshall EJP (1993) Exploiting seminatural habitats as part of good agricultural practice. Scientific basis for codes of good agricultural practice. Jordan WL (ed) EUR 14957. Commission for the European Communities, Luxembourg

Mas JF, Velázquez A, Díaz-Gallegos JR, Mayorga-Saucedo R, Alcántara C, Bocco G, Castro R, Fernández T, Pérez-Vega A (2004) Assessing land use/cover changes: a nationwide multidate spatial database for Mexico. Int J Appl Earth Obs Geoinf 5(4):249–261

McCarter J, Gavin MC (2014) Local perceptions of changes in traditional ecological knowledge: a case study from Malekula Island, Vanuatu. Ambio 43(3):288–296

Nautiyal S, Kaechele H (2008) Management of: environmental. Manag Environ Qual 19(3)

Noordijk J, Musters CJM, van Dijk J, de Snoo GR (2010) Invertebrates in field margins: taxonomic group diversity and functional group abundance in relation to age. Biodivers Conserv 19(11):3255–3268

Pandey DN (2002) Traditional knowledge systems for biodiversity conservation. Organization of the United Nations (FAO) Forestry Paper. FAO Rome Italy 22–41

Purohit A, Maikhuri RK, Rao KS, Nautiyal S (2002) Revitalizing drink: an assessment of traditional knowledge system in Bhotiya community of Central Himalayas, India

Ramachandra TV (2012) Peri-urban to urban landscape patterns elucidation through spatial metrics. Int J Eng Res Technol 2(12):58–81

Ramakrishnan PS (2002) Coping with uncertainties: role of traditional ecological knowledge in socio-ecological landscape management. Proc Indian Natl Sci Acad Part-B 68(3):235–254

Ramakrishnan PS (2003) Linking natural resource management with sustainable development of traditional mountain societies. J Trop Ecol 44(1):43–54

Ramakrishnan PS, Das AK, Saxena KG (1996) Conserving biodiversity for sustainable development. Indian National Science Academy, New Delhi

Rist LR, Uma Shaanker EJ, Milner-Gulland GJ (2010) The use of traditional ecological knowledge in forest management: an example from India. Ecol Soc 15(1):3

Röös E, Mie A, Wivstad M, Salomon E, Johansson B, Gunnarsson S, Wallenbeck A, Hoffmann R, Nilsson U, Sundberg C, Watson CA (2018) Risks and opportunities of increasing yields in organic farming a Review. Agron Sustain Dev 38:14

Schmitz J, Hahn M, Brühl CA (2014) Agrochemicals in field margins—an experimental field study to assess the impacts of pesticides and fertilizers on a natural plant community. Agric Ecosyst Environ 193:60–69

Sitzia T, Campagnaro T, McCollin D, Dainese M (2012) A miscellany of traditional management techniques of woody field margins on the Po Plain, Italy: implications for biodiversity conservation. In: Dover JW (ed) Hedgerow futures. Hedgelink, Stoke-on-Trent, UK, pp 135–146

Sparks TH, Parish T (1995) Factors affecting the abundance of butterflies in field boundaries in Swavesey fens, Cambridgeshire, U.K. Biol Cons 73:221–227

Swamy RN (2014) Protection of traditional knowledge in the present IPR regime: a mirage or a reality. Indian J Public Adm 60(1):35–60

Voeks RA, Leony A (2004) Forgetting the forest: Assessing medicinal plant erosion in Eastern Brazil. Econ Bot 58:S294–S306

Wood BM (2013) Do legume-rich habitats provide improved farmland biodiversity resources and services in arable farmland? Ann Appl Biol 118:239–246

Woodcock BA (2005) Pitfall trapping in ecological studies. Insect Sampling Forest Ecosyst 37–57

Printed in the United States
by Baker & Taylor Publisher Services